T0073169

Come On!

Ernst Ulrich von Weizsäcker • Anders Wijkman

Come On!

Capitalism, Short-termism, Population and the Destruction of the Planet

A Report to the Club of Rome

by Ernst von Weizsäcker and Anders Wijkman, co-authors in cooperation with 34 more Members of the Club of Rome
prepared for the Club of Rome's 50th Anniversary in 2018

Authors: Ernst Ulrich von Weizsäcker and Anders Wijkman, Co-Presidents, The Club of Rome.

Contributors (alphabetical order): *Carlos Alvarez Pereira, Nora Bateson, Mariana Bozesan, Susana Chacón, Yi Heng Cheng, Robert Costanza, Herman Daly*, Holly Dressel, Lars Engelhard, *Herbie Girardet, Maja Göpel, Heitor Gurgulino de Souza, Karlson "Charlie" Hargroves, Yoshitsugu Hayashi, Hans Herren, Kerryn Higgs, Garry Jacobs*, Volker Jäger, *Ashok Khosla*, Gerhard Knies, *David Korten, David Krieger, Ida Kubiszewski, Petra Künkel, Alexander Likhotal,* Ulrich Loening, *Hunter Lovins, Graeme Maxton, Gunter Pauli, Roberto Peccei, Mamphela Ramphele, Jørgen Randers, Kate Raworth, Alfred Ritter, Joan Rosàs Xicota, Peter Victor, Agni Vlavianos Arvanitis* and *Mathis Wackernagel* (Club of Rome members in italics).

Ernst Ulrich von Weizsäcker
Emmendingen, Germany

Anders Wijkman
Stockholm, Sweden

ISBN 978-1-4939-7418-4 ISBN 978-1-4939-7419-1 (eBook)
DOI 10.1007/978-1-4939-7419-1

Library of Congress Control Number: 2017952604

© Springer Science+Business Media LLC 2018
This work is subject to copyright. All rights are reserved by the Publisher, whether the whole or part of the material is concerned, specifically the rights of translation, reprinting, reuse of illustrations, recitation, broadcasting, reproduction on microfilms or in any other physical way, and transmission or information storage and retrieval, electronic adaptation, computer software, or by similar or dissimilar methodology now known or hereafter developed.
The use of general descriptive names, registered names, trademarks, service marks, etc. in this publication does not imply, even in the absence of a specific statement, that such names are exempt from the relevant protective laws and regulations and therefore free for general use.
The publisher, the authors and the editors are safe to assume that the advice and information in this book are believed to be true and accurate at the date of publication. Neither the publisher nor the authors or the editors give a warranty, express or implied, with respect to the material contained herein or for any errors or omissions that may have been made. The publisher remains neutral with regard to jurisdictional claims in published maps and institutional affiliations.

Printed on acid-free paper

This Springer imprint is published by Springer Nature
The registered company is Springer Science+Business Media LLC
The registered company address is: 233 Spring Street, New York, NY 10013, U.S.A.

Preface

In the years since its founding in 1968, there have been more than 40 reports to the Club of Rome. The first report, *The Limits to Growth*, catapulted the Club of Rome, and the authors of *Limits*, into the global limelight. The book served as a shock to a world as yet largely unaware of the long-term effects of continued growth in what we now call the human ecological footprint. Aurelio Peccei, founder and then president of the Club of Rome, saw the responsibility of addressing the suite of problems facing the world, what he called the predicament of mankind, but was astonished to learn from the *Limits* report that these problems could all be tied to the consequences of humankind's desire for endless growth on a finite planet. The message from the bold young team at the Massachusetts Institute of Technology was that if growth continued unabated at the present pace, shrinking resources and heavy pollution would lead to an ultimate collapse of world systems.

Certainly, today's computer models are much more advanced than the World3 model used by the 1972 team. Some aspects of economic growth during the almost five decades that have passed – like innovation – were not fully taken into account. But the central message of *Limits* is as valid today as it was in 1972. The world of today is facing many of the challenges that were anticipated in the 1970s: climate change, scarcity of fertile soils, and massive species extinction. Furthermore, the planetary social situation remains extremely unsatisfactory, with some four billion people living in very tenuous economic conditions or being threatened by natural disasters or wars. New estimates warn that more than 50 million people will be forced every year to leave their home and emigrate. Where can they go? In 2017, there are already 60 million refugees in the world!

Simultaneously, however, modern societies have acquired an amount of economic wealth, scientific knowledge, and technological capacities that should enable to fund and implement most of the transformations that *The Limits to Growth* saw as paramount in terms of creating a sustainable world.

We, the Executive Committee of the Club of Rome, gratefully acknowledge the merits and message of *The Limits to Growth*, as well as that of the other very valuable reports that have been written to the Club of Rome. Moreover, we remember the bold step taken in 1991 by Alexander King, Aurelio Peccei's successor as president of the Club of Rome, who published *The First Global Revolution*, a book, coauthored by Bertrand Schneider, then the club's secretary general. In contrast to other reports, *The First Global Revolution* was presented as a report by the Council of the Club of Rome (the equivalent to today's Executive Committee of the Club of Rome). King and Schneider realized that the end of the Cold War opened huge new opportunities that could lead to a peaceful and prospering world. This optimistic book brought the Club of Rome back into the limelight, albeit less so than had *The Limits to Growth*.

The world is again in a critical situation. We see the need for a bold new beginning. This time, however, we believe it is particularly important to look at the philosophical roots of the current state of the world. We must question the legitimacy of the ethos of materialistic selfishness that is currently the most powerful driving force in the world, and we welcome Pope Francis's initiative in addressing a deeper-lying crisis of values, a central issue which the Club of Rome identified many years ago. The time has come, we believe, for a new Enlightenment or for otherwise overturning current habits of thought and action that only consider the short term. We acknowledge the strong approach taken by the United Nations in their 2015 formulation of the 2030 Agenda, comprising 17 Sustainable Development Goals to be implemented over the next 15 years. However, unless the destructive driving forces of purely materialistic economic growth are tamed, we cannot escape the fear that 15 years from now the world will be in an even harsher ecological situation than it is today.

From this perspective, the committee wholeheartedly supports the initiative taken by our current copresidents in composing and coordinating a new and ambitious report that addresses the predicament of humankind from the perspective of today's realities.

And now, a word of explanation for the surprising title. "Come on" has several different meanings in the English language. In casual language, it is often spelled "C'mon" and means "don't try to fool me." We consider this the meaning for Chaps. 1 and 2 of the book. We don't want to be fooled by the usual descriptions of the state of the world and the usual, corresponding answers, which can make things worse, not better. And we don't want to be fooled by outdated philosophies. Another meaning of the title is thoroughly optimistic: "Come on, join us!" This is the meaning for Chap. 3 of the book, which we consider an exciting journey of real solutions. Clearly, the architecture of the book comprises both meanings but in the indicated order. (To be sure, also some more meanings, including somewhat dirty ones, of "Come on" exist, but they have no relevance for us!)

June 2017. The Executive Committee of the Club of Rome.

Susana Chacón, Enrico Giovannini, Alexander Likhotal, Hunter L. Lovins, Graeme Maxton, Sheila Murray, Roberto Peccei, Jørgen Randers, Reto Ringger, Joan Rosàs Xicota, Ernst von Weizsäcker, Anders Wijkman, and Ricardo Díez Hochleitner (Honorary Member).

Executive Summary

The human-dominated world can still have a prosperous future for all. This requires making sure that we do not continue to degrade our planet. We firmly believe this is possible, but it becomes increasingly difficult to achieve, the longer we wait to act appropriately. Current trends are in no way sustainable. Continued conventional growth leads to massive collisions with natural planetary boundaries. The economy under the dictates of the financial system with its seduction to speculation tends to lead to widening gaps in terms of wealth and income.

World population must be stabilized soon, not just for environmental but also for compelling social and economic reasons. Many people see the world in a state of disarray, confusion, and uncertainty. Deep social inequalities, failed states, wars and civil wars, unemployment, and mass migrations have left hundreds of millions of people in a state of fear and despair.

The United Nations has unanimously adopted the *2030 Agenda*, which is meant to address these challenges. However, a successful implementation of the agenda's 11 socioeconomic goals could more than likely destroy its three ecological goals, which are to stabilize the climate, restore the oceans, and halt biodiversity loss. The only way to avoid this to happen would be by adopting an integrated approach to policymaking, leaving behind today's silo-based structures.

Chapter 1 of this book offers a diagnosis of the non-sustainable trends of our time, of what has been termed the "Anthropocene" – the age of human domination of all aspects of this planet, including its biogeochemical composition. A "prosperous future for all" requires that economic well-being be largely decoupled from the destruction of natural resources, especially in agriculture, and the pollution of the atmosphere. The book suggests that the legitimacy of full national sovereignty must be questioned concerning all matters that affect the entire globe.

Chapter 2 offers a deeper analysis, describing society's fundamental philosophical crisis at this juncture, starting with the encyclical letter *Laudato Sí* by Pope Francis. The foundations of today's religions and common beliefs, as well as our system of economics, stem from a time of the "empty world" (Herman Daly) and are inappropriate for our current "full world." Capitalism as we know it, with its focus on short-term profit maximization, is moving us in the wrong direction – towards an increasingly destabilized climate and degraded ecosystems. In spite of all the knowledge we have today, we seem unable to change course, literally driving planet Earth to destruction. Ultimately, Chap. 2 suggests the need of a new Enlightenment, one that is fitting for the "full world" and for *sustainable* development. That enlightenment should embrace the virtues of *balance* instead of doctrine. We explicitly mention the balance between humans and nature, between short term and long term, and between public and private interests. Chapter 2 can be seen as the most revolutionary part of the book.

Can the planet's beleaguered natural systems wait until all of human civilization has gone through the long process of a new Enlightenment? No, explains Chap. 3; we must act now. This is absolutely doable. We list an optimistic if slightly haphazard collection of opportunities that already exist: decentralized clean energy, sustainable jobs in every type of country, and a massive decoupling of human well-being from the use of fossil fuels, basic materials, and scarce minerals. Pragmatic policies including on the financial system are featured. Frame conditions must make sustainable technologies truly profitable and encourage investors to support long-term solutions.

The book closes with an invitation to readers and discussants to engage themselves in the many possible ways of creating a sustainable world society.

Acknowledgments

This report is a multi-contributor book. We as lead authors very gratefully acknowledge the excellent contributions received, in draft form, from Nora Bateson (parts of Sect. 2.7), Mariana Bozesan (Sect. 3.13), Yi Heng Cheng (Sect. 3.17), Herman Daly (Sects. 1.12 and 2.6.2), Lars Engelhard (parts of Sect. 3.13), Herbie Girardet (Sects. 1.7.2 and 3.6), Maja Göpel (Sect. 1.1 and linking sections between the three chapters), Garry Jacobs and Heitor Gurgulino de Souza (Sects. 2.8 and 3.18), Volker Jäger and Christian Felber (Sect. 3.12.4) Karlson "Charlie" Hargroves (Sect. 3.9), Yoshitsugu Hayashi (Sect. 3.6.3), Hans Herren (Sects. 1.8 and 3.5), Kerryn Higgs (Sects. 1.9 and 3.11 and several other pieces), Ashok Khosla (Sect. 3.2), Gerhard Knies (Sect. 3.16.3), David Korten (Sect. 2.2), David Krieger (Sect. 1.6.2), Ida Kubiszewski and Robert Costanza (Sect. 3.14 and part of Sect. 1.12), Petra Künkel (Sect. 3.15), Ulrich Loening (essential comments on Sects. 2.6 and 2.7), Hunter Lovins (Sect. 3.1 and parts of Sects. 1.6 and 3.4), Graeme Maxton (Sects. 2.5 and 3.12.2), Gunter Pauli (Sect. 3.3), Roberto Peccei (Preface, Chap. 1, and structure), Jørgen Randers (Sects. 2.5 and 3.12.2), Kate Raworth (Sect. 3.12.1), Alfred Ritter (part of Sect. 3.5), Joan Rosàs Xicota (essential comments on Sects. 1.1.2 and 3.11), Agni Vlavianos Arvanitis (part of Sect. 3.6), and Mathis Wackernagel (part of Sect. 1.10). In all cases, we as authors applied modifications with a view of making the book coherent in substance and style. But without the valuable contributions, we would have been at a loss.

Kerryn Higgs, Mamphela Ramphele, Jørgen Randers, Alexander Likhotal, Ulrich Loening, David Korten, Irene Schöne, Mathis Wackernagel, and Jakob von Weizsäcker went through the trouble of looking at the entire manuscript or at least major parts of it and greatly helped us discovering weaknesses and omissions. Susana Chacón and Peter Victor made very important oral comments during a preparatory meeting on the book in May 2016. Verena Hermelingmeier accompanied the authors throughout the process of writing and helped in formulating major passages. Hans Kretschmer oversaw the quality of illustrations and where necessary secured printing permissions for them.

Toward the end, we engaged Holly Dressel as the main language editor for the entire book. It turned out she did a lot more than language editors do. She became a true contributor making the text a lot more readable and attractive, offering new phrases, and finding good references.

As authors, we are truly grateful to the members of the Executive Committee of the Club of Rome for accompanying and encouraging us in writing this book.

We are also very grateful to our club member Alfred Ritter for supporting the initiative of writing this ambitious work and for financing it to a large extent. We gratefully acknowledge additional financial support by the Robert Bosch Foundation.

Emmendingen, Germany, and Stockholm, Sweden, June 2017

Ernst von Weizsäcker and Anders Wijkman, Copresidents of the Club of Rome

Contents

Chapter 1
C'mon! Don't Tell Me the Current Trends Are Sustainable!

1.1 Introduction: The World in Disarray

We all know that the world is in crisis. Science tells us that almost half of the top soils on earth have been depleted in the last 150 years[1]; nearly 90% of fish stocks are either overfished or fully fished.[2] Climate stability is in real danger (Sects. 1.5 and 3.7); and the earth is now in the sixth mass extinction period in history.[3]

Perhaps the most accurate account of the ecological situation is the 2012 'Imperative to act',[4] launched by all the 18 recipients (till 2012) of the Blue Planet Prize, including Gro Harlem Brundtland, James Hansen, Amory Lovins, James Lovelock and Susan Solomon. Its key message reads, 'The human ability to do has vastly outstripped the ability to understand. As a result, civilization is faced with a perfect storm of problems, driven by overpopulation, overconsumption by the rich, the use of environmentally malign technologies and gross inequalities'. And further, 'The rapidly deteriorating biophysical situation is barely recognized by a global society infected by the irrational belief that physical economies can grow forever'.

[1] Arsenault (2014).

[2] FAO (2016).

[3] Kolbert (2014).

[4] Blue Planet Prize Laureates (2012).

© Springer Science+Business Media LLC 2018

E.U. von Weizsäcker, A. Wijkman, *Come On!*,
DOI 10.1007/978-1-4939-7419-1_1

1.1.1 Different Types of Crisis and a Feeling of Helplessness

The crisis is not cyclical but growing. And it is not limited to the nature around us. There are also a social crisis, a political and a cultural crisis, a moral crisis, as well as a crisis of democracy, of ideologies and of the capitalist system. The crisis also consists of deepened poverty in many countries and the loss of jobs for a considerable part of the population worldwide. Billions of people have reached a state of mind where they don't trust their government anymore.[5]

Seen from a geographic point of view, symptoms of crisis are found nearly everywhere. The 'Arab Spring' was followed by a series of wars and civil wars, serious human rights violations and many millions of refugees. The internal situation is not better in Eritrea, South Sudan, Somalia, Yemen or Honduras. Venezuela and Argentina, once among the richer states of the world, face huge economic challenges, and neighbouring Brazil has gone through many years of recession and political turmoil. Russia and several East European countries are struggling with major economic and political problems in their post-communist phase. Japan finds it difficult to overcome decade-long stagnation, and to deal with the 2011 tsunami and ensuing nuclear disaster. And the temporary economic upswing several African countries have enjoyed lost its dynamism as soon as the prices of mineral resources collapsed, and partly due to very unusual droughts. Land grabbing is plaguing much of Africa, but also other parts of the world, leading to involuntary dislocations of millions of people and the related problems with refugees both within countries and abroad.[6]

The response of governments has been concentrated, at worst, on managing their own political image, and at best to treat the symptoms of the crisis, not the cause. The problem is that the political class in the whole world is strongly influenced by investors and by powerful private companies.

This indicates that the current crisis is also a crisis of global capitalism. Since the 1980s, capitalism has moved from furthering the economic development of countries, regions and the world towards maximizing profits, and then to a large extent profits from speculation. In addition, the capitalism unleashed since 1980 in the Anglo-Saxon world, and since 1990 worldwide, is mainly financial. This trend was supported by excessive deregulation and liberalization of the economy (see Sect. 2.4). The term 'shareholder value' popped up in the business pages of the media worldwide, as if that was now the new epiphany and guardrail for all economic action. In reality, it served to narrow business down to short-term gains, often at the expense of social and ecological values. The myth of shareholder value has been effectively debunked in a recent book by Lynn Stout.[7]

A different, if related, feature of 'disarray' is the rise of aggressive, mostly right-wing movements against globalization in OECD countries, often referred to as populism. These have become overt through Brexit and the Trump victory in the United States. As Fareed Zakaria observes, 'Trump is part of a broad populist

---

[5] The Edelman Trust Barometer (2017) says that 53% of the population in 28 countries believe the systems governing them are failing; only 15% deem that the systems are working.

[6] Liberti (2013).

[7] Stout (2012).

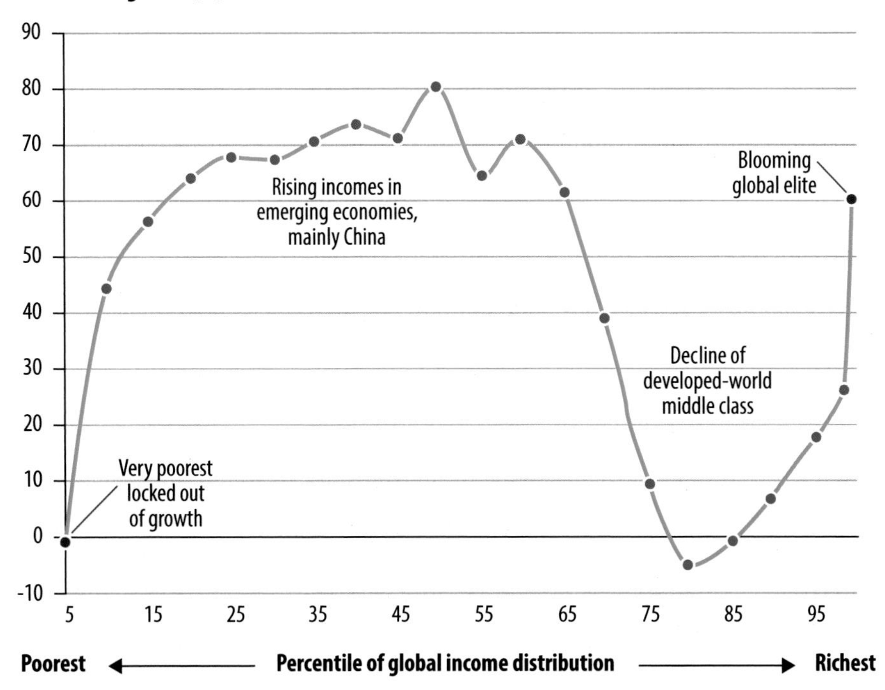

Fig. 1.1 Global income growth from 1988 to 2008 for 21 income groups from poorest to richest. The curve resembles the silhouette of an elephant and is referred to as 'elephant curve' (Source: http://prospect.org/article/worlds-inequality)

upsurge running through the Western world. … In most (countries), populism remains an opposition movement, although one that is growing in strength; in others, such as Hungary, it is now the reigning ideology'.[8]

This phenomenon of right-wing populism can be explained to an extent by the 'trunk valley of the elephant curve' (Fig. 1.1)[9] showing the decline of *developed-world* middle classes, during a 20-year period. While more than half of the world's population was enjoying over 60% income rises, OECD's middle classes suffered losses caused mainly by the deindustrialization and job losses in major parts of the United States, Britain and other countries. In the United States, the median income increased by a meagre 1.2% since 1979.

The stunning income growth on the left-hand side of the curve, the 'back of the elephant', lifting some two billion people out of poverty, was caused mainly by China's and some other countries' economic success. What remains invisible on the picture is the far end of 'the trunk of the elephant': The richest 1% of the world and, more revolting, the richest eight persons of the world now own as much wealth as

[8] Zacharia (2016).

[9] BrankoMilanovic.2016.https://milescorak.com/2016/05/18/the-winners-and-losers-of-globalization-branko-milanovics-new-book-on-inequality-answers-two-important-questions/

the poorest half of the world population combined, a figure publicized by Oxfam during the 2017 World Economic Forum.[10]

The 'elephant curve' gives an incomplete picture for a second reason. The Oxford Poverty and Human Development Initiative (OPHI) has proposed a Multidimensional Poverty Index (MPI) going beyond just income and including ten indicators around health, education and living standards. Using that MPI, OPHI counts 1.6 billion people living in 'multidimensional poverty' in 2016 – nearly *twice* as many as the number of people living in extreme poverty measured by income alone.[11]

Thirdly, the interpretation of the curve requires an analysis of the people in each percentile group. In fact, they tend to move. And the curve does not distinguish those in Russia and East European countries who lost much of their income after 1990 from those in Detroit or middle England who, for very different reasons, also were among the losers.[12] Another fact cannot be seen in the picture: the massive shift of money and income from the manufacturing and trade sectors to the financial sector.[13] Bruce Bartlett, a senior policy advisor to both the Reagan and Bush administrations, argues that this 'financialization' of the economy is the cause of income inequality, falling wages and the poor performance. David Stockman, Reagan's director of the Office of Management and Budget, agrees, describing our current situation as 'corrosive financialization that has turned the economy into a giant casino since the 1970s'.[14]

Populist politicians in the OECD countries see themselves as speaking for the forgotten 'ordinary' people and for genuine patriotism, but they tend to fight and antagonize the people representing *democratic* institutions – what an irony!

For the European Union (EU), the strongest trigger for populism has been the millions of refugees who came or would like to come to Europe from the Near East, from Afghanistan and from Africa. Even the most generous European countries have reached their own assumed limits for receiving these masses of refugees. The EU institutions were too weak (not too powerful, as they are depicted by the new nationalists) to deal with the 'refugee crisis', resulting eventually in an identity crisis in the EU. Once a success story of an entity ensuring peace and economic development, the EU has lost some of its unifying narrative. The populist right-wing movements or parties see and criticize the EU as the culprit for all kinds of undesired events. The irony is that continuing the success story would require more, not less, powers for the Union. The Union should be entrusted with border protection, a well-funded common asylum and refugee policy to deal with the refugee crisis and maintain the advantages of the Schengen agreement. And for the re-stabilization of the Euro, the EU or at least the Euro zone needs a common fiscal policy, as the new French President Emmanuel

[10] https://www.oxfam.org. 2017-01-16. Just eight men own same wealth as half the world. The title of the study is "An economy for the 99 percent." Data are based on the *Credit Suisse Global Wealth Data* book, 2016. See also Jamaldeen (2016).

[11] OPHI (2017). See also Dugarova and Gülasan (2017).

[12] For more details see Corlett (2016).

[13] Greenwood and Scharfstein (2013). Authors say that in 1980, people working in the financial sector made about the same as people in other industries; By 2006 they made 70% more.

[14] Bartlett (2013). Stockman (2013).

Macron is proposing. But it is these very measures of which nationalist populists are most afraid.

The EU in its present form is not without shortcomings. Free market principles have come to dominate EU policymaking, leading to a subordination of other policies, like environment. Notably the UK wanted that priority, as it preferred to see the EU chiefly as a union for mutual trade. And the austerity policies pursued have blocked many benign investments and led to unnecessary suffering among tens of millions of Europeans. Such shortcomings, however, should never be used to put in question the overall objectives of the EU – a union of peace, the rule of law, human rights, cultural understanding and sustainability.

Addressing the global crisis of democracy, the German Bertelsmann Foundation has published a 3000-page empirical report on progress (or lack thereof) on democracy and a social market economy, as measured by the Bertelsmann Transformation Index (BTI).[15] Over the last few years, the report sees a consistent decay of such parameters as civil rights, free and fair elections, freedom of opinion and of press, freedom of assembly and separation of powers. Within the same time frame, the number of countries in which authoritarian, mostly religious, dogmas influence political decision making rose from 22% to 33%. That report was published before the assaults on democracy and civil rights that occurred in summer 2016 in Turkey or the Philippines. Symptoms of tyranny are spreading, including in some of the countries with a solid tradition of freedom and democracy.[16]

Let us briefly turn to a different kind of crisis. Well, not exactly a crisis but an unpleasant feature in an otherwise fruitful communication tool, the 'social media'. Aside from being practical and useful for everyday arrangements and exchange of news and reasonable opinions, social media also have become vehicles for enhancing conflicts and vilification of mostly innocent individuals, and for spreading 'post truth' nonsense. Much of the contents of social media political conversation is self-enhancing political rubbish, as those media serve as 'echo chambers' for networks of like-minded frustrated citizens.[17] An empirical study from China found that anger and indignation are the emotions that are most likely to get viral in the social media, meaning they are multiplied faster and stronger than other emotions.[18]

The Internet and the social media are also vehicles for 'bots' (short for robots) that can disrupt or destroy messages, multiply nonsense and create all kinds of mischief. There are dozens of types of malicious bots (and botnets) to harvest email addresses, to grab content of websites and reuse it without permission, to spread viruses and worms, to buy up good seats for entertainment events, to increase views for YouTube videos or to increase traffic counts in order to extract money from advertisers.

A more frightening cause of disarray relates to terrorism. In earlier times, humanity's violent conflicts occurred mostly between different countries. In recent times,

[15] Bertelsmann Stiftung. 2016. (Lead author: Sabine Donner) Politische und soziale Spannungen nehmen weltweit zu. Executive Summary. Transformationsindex der Bertelsmann Stiftung. Gütersloh.

[16] Snyder (2017).

[17] Quattrociocchi et al. (2016).

[18] Fan et al. (2014).

systemic and at least partly religious conflicts prevail, using terror attacks with the explicit intention of making people feel insecure. During much of the twentieth century, religions remained quiet, non-aggressive and geographically confined to rather stable territories. This no longer is true. Partly because of globalized populations moving or being forced to leave their home territories, some factions of Islam have expanded geographically and are claiming strong influence over national states, for example, attacking countries like France with its tradition of laicism that does not permit religion to dominate politics.

What tends to be underrepresented in the media is the *positive* role of religions. In Christian-dominated Europe, liberal and tolerant religion became part of the European identity a century after the Enlightenment successfully discredited the earlier doctrinaire, authoritarian and colonialist-missionary manifestations of the faith. During the Cold War, Christian goals of social cohesion helped build the system of 'Western values', often described as the social welfare state, or the 'social market economy' (for its partial demise, see Sect. 2.4).

With a view towards leading Islam into an equally benign and co-operative social role, some Islamic scholars, such as Syrian born Bassam Tibi, call on Muslims in Europe to integrate into democratic society.[19] Tibi, however, is not popular among radical Muslims, to put it mildly. But to understand the radicalization of Islam, one must not underestimate the role played by the West, in particular the United States, in interfering with Near Eastern states.

Some would say that the troublesome situations mentioned so far, the recurring topics of media headlines, are only the surface of our world's 'disarray'. Deeper and more systemic problems include the breath-taking speed of technological development that may very easily run out of control. One trend is digitization that potentially threatens millions of jobs (see Sect. 1.11.4). Another trend or development can be observed in the biological sciences and technologies. The enormous acceleration of genetic engineering through the CRISPR-Cas9 technology[20] is causing fears of monster creation or the extinction of species or varieties not seen as valuable under human utilitarian criteria. Generally, a non-specific feeling is spreading that 'progress' has scary sides and that the genie may already have left the bottle (see Sect. 1.11.3).

No doubt there is a need to analyse and understand the symptoms and roots of the variety of crises, political, economic, social, technological and environmental. It is also important to recognize the extent to which people perceive the various phenomena of disarray and feel disoriented, and to recognize that the reality and the feelings of disarray have a moral and even religious dimension.

[19] Tibi (2012). He sees "Islamism" as incompatible with democracy, while Islam has deep roots into democratic consultation methods and has been open for a very early Enlightenment in the twelfth Century, chiefly through Ibn Rushd – Latinised as Averroes.

[20] E.g. Hsu et al. (2014).

1.1.2 Financialization: A Phenomenon of Disarray

An important part of the disorientation relates to financial markets. Historians will look back at the last 30 years with concern, when looking at the explosion in bank balance sheets, backed up by declining levels of equity and massive borrowing. One of the results was a temporary private-sector-led boom. The other was a massive increase in the world's financial sector (finance, insurance, real estate – FIRE), often called *financialization*, and subsequently the financial crisis of 2008–2009.

Excessive risk-taking developed into a crisis that was close to bringing the whole financial system to a halt. When the bubble burst, many governments were forced to step in with broad support programmes.

Governments caught by the new mind-set (see Sect. 2.4) were intimately involved in all of this. True, there are many examples of serious malpractices within the private financial sector. But had it not been for the systematic deregulation of the banks by governments, with the purpose of stimulating economic growth by issuing more debt, the situation would have been radically different. The causes behind the crisis were many and varied:

- Excessive lending by the banking industry
- Lack of action on the part of regulators and central banks to stop (i) excessive lending, (ii) the spread of exotic financial instruments (synthetic assets and bonds, collateralized mortgage obligations/CMOs, structured debt issues, etc.) and (iii) pure speculative transactions
- Opaque tax havens, and the absence of a binding legal framework that is accepted and implemented by the international community, in general, and the major jurisdictions and financial centres
- Securitization and distribution by investment banks and other financial actors of mortgage-related assets and investment vehicles transferring the credit risk from the original lender to the ultimate bondholders
- Failure by some rating agencies and auditing firms to properly assess and report the inherent risks posed by many of the financial products

A deeper analysis is presented by economists Anat Admati and Martin Hellwig[21] about the main causes behind the financial crisis. Western banks borrowed far too much with far too little equity in their balance sheets to act as a buffer if things went wrong in their business – from trading in the multitrillion-dollar derivatives markets to often reckless lending on real estate. In the decades following the Second World War, banks operated with between 20% and 30% of their liabilities as equity. By 2008, that had shrunk to just 3%. Banks obviously believed that they had invented instruments that removed the risk, allowing them to run their banks with a tenth of the buffer they had before. It proved to be very unrealistic. But they counted with the state to underwrite their risks.

[21] Admati and Hellwig (2013).

Bankers have enriched themselves spectacularly in the process. They made themselves 'too big to fail' – and too big to jail. The 2008 financial crisis was mostly caused by that irresponsible greed.[22] Yet, in 2009, not only did bankers avoid criminal prosecutions and receive hundreds of billions in government bailouts, but some still paid themselves record bonuses. At the same time, almost nine million households in the United States had to abandon their homes when the value of their houses plummeted and they could no longer service the adjustable-rate mortgages – the so-called foreclosure crisis.[23]

Financialization refers to the dominance of the financial sector in the global economy and the tendency for accumulated profits (and leverage) to flow into real estate and other speculative investment. Debt is an intrinsic element in this process. In the United States, for example, both household debt and private sector debt more than doubled relative to GDP between 1980 and 2007.[24] The same is true for most OECD countries. At the same time, 'the value of financial assets grew from four times GDP in 1980 to ten times GDP in 2007 and the finance sector's share of corporate profits grew from about 10% in the early 1980s to almost 40% by 2006'.[25] Adair Turner, chair of the UK's Financial Services Authority in the years following the 2007–2008 crisis, regards unchecked private credit creation as the key system fault that led to that crisis with its devastating consequences.[26] From this follows that the financial sector constitutes a significant and increasing risk factor in the economy.

The degree of financialization varies from country to country but the increase in the power of finance is general. The current finance sector evolved in the context of the deregulation that gathered pace from the late 1970s and expanded dramatically after the 1999 removal of the separation between commercial and investment banking in the United States.[27] This barrier had been put in place in 1933 by the Roosevelt administration in response to the Wall Street Crash of 1929, when a period of rampant credit creation and financial speculation collapsed. Similar speculation preceded the crisis of 2007–2008: The face value of financial products reached US$640 trillion in September 2008, 14 times the GDP of all the countries on earth.[28]

Lietaer et al.[29] compare speculation with ordinary money transfers paying for goods and services: 'In 2010, the volume of foreign exchange transactions reached $4 trillion per day', which does not even include derivatives. In comparison, 'one day's exports or imports of all goods and services in the world amount to about 2% of those $4 trillion'. Transactions *not* paying for goods and services, almost by definition are

---

[22] E.g. McLean and Nocera (2010).

[23] NCPA (2015).

[24] "In 1981 household debt was 48% of GDP, while in 2007 it was 100%. Private sector debt was 123% of GDP in 1981 and 290% by late 2008" (Crotty 2009, p. 576).

[25] Crotty (2009), ibid.

[26] Turner (2016). "Across advanced economies private-sector debt increased from 50% of national income in 1950 to 170% in 2006"(p. 1).

[27] The removal of the separation occurred in 1986 in the UK.

[28] Sassen (2009).

[29] Lietaer et al. (2012). Quotes from pages 11–12.

speculative. Such financial products and transactions, the authors continue, lead regularly to monetary crashes, sovereign debt crises and systemic crashes with an average of more than ten countries in crisis every year.

One of the consequences of this development is that a significant part of economic growth has been distributed to the wealthy, as mentioned with the new Oxfam figures in the previous subchapter.

Practices within the financial sector demonstrate a disregard for the impact they have on both people and the planet. That includes a distinct short-termism, the ratio of banks' reserves to their loans, the ratio of banks' lending that support the real economy versus speculation in property and derivatives, unchecked credit creation – in fact money creation – and the failure to account for long-term climate and environmental risks. In the words of Otto Scharmer at MIT,[30] 'We have a system that accumulates oversupply of money in areas that produce high financial and low environmental and social returns, while at the same an undersupply of money in areas that serve important societal investment needs'.

The failure to account for environmental risks means that the pressure on already-scarce natural resources accelerates – trees are felled, waterways polluted, wetlands drained and the exploitation of oil, gas and coal accelerating, as long as there is demand. It also means that huge savings, among them pension funds, are locked into investments in fossil-based assets. Such assets are increasingly looked upon as high-risk assets (see Sect. 3.4).

1.1.3 Empty World Versus Full World

The Club of Rome was always conscious of the philosophical roots of human history. Among the valuable scripts are Kenneth Boulding's *The Meaning of the Twentieth Century* saying (in short) that the meaning is the stewardship of Spaceship Earth. His book was labelled one of the five 'prescient classics that first made sustainability a public issue'.[31]

But then many thinkers saw that the stewardship was difficult under the conditions of the *full world*.[32] That became the chief message of the Club of Rome during its early years, written down in *The Limits to Growth*.[33] Humans cannot become successful stewards of Spaceship Earth with development ideals, scientific models and value sets that were shaped at a time of the *empty world*, when the population was small and the bounty of natural resources on this earth seemed endless, that is, during the time when the European Enlightenment unfolded and the Americas looked like places where settlers and entrepreneurs could endlessly find new space.

---

[30] Scharmer (2009).

[31] Rome (2015).

[32] Daly (2005); see also Sect. 1.12.

[33] Meadows et al. (1972).

Today, actually since the mid of the twentieth century, humanity lives in a *full world*. The limits are tangible, palpable in almost everything people do. And yet, 45 years after *The Limits to Growth* became a public issue, the world still tracks along the 'standard run' of the 1972 *Limits* model, representing the business as usual development stemming from the *empty world*. Recent studies[34] actually support the *Limits*' predictive relevance. A new term illustrating the limits phenomenon is that of the *planetary boundaries*[35] (see Sect. 1.3).

When *Limits* was published, many people, notably in the political domain, feared the message was that humanity had to give up on prosperity and agreeable life styles. But that was never the idea of the Club of Rome. Our main concern was the growing footprint of mankind and that economic activity has to assume radically different forms.

Why it is so hard to change the old trends? Well, changing trends depends on changing minds. That was the experience of the European Enlightenment. That courageous process took roughly two centuries, the seventeenth and eighteenth centuries, and served as a great liberation from authoritarian rules and narratives defined by the Crown and the Church. The enlightened transformation was successful because it championed human reasoning and rational change through the application of the scientific method. The Enlightenment established the ideals of individual freedom, economic growth and technological innovation that had barely existed previously in European society. The concepts of democracy and the separation of powers brought political influence to many more men (hardly women) or their elected representatives. And innovators, entrepreneurs and merchants were allowed to flourish and to become a new 'aristocracy', this time legitimated by their own work, not by royal families. The Enlightenment was experienced by most people in Europe as an extremely welcome development.

There were dark sides too. European colonialism with all its arrogance and cruelty found almost no critique among the intellectuals of the Enlightenment. The misery of the working classes and impoverished peasants to say nothing of the colonized indigenous peoples all over the world was hardly noticed in bourgeois circles. No comprehension was visible of the equivalent value of women and men. And unrestrained growth was seen as completely legitimate.

History continues. Global population rose from one billion in the eighteenth century to some 7.6 billion today. In parallel, per capita consumption of energy, water, space and minerals was also growing. This twin development catapulted us into the 'full world'. Looking at ecological and economic realities, the time has come for demanding some kind of new Enlightenment, one that fits for the full world. Growth may no longer be automatically related to living better lives, but can actually be detrimental. This simple but fundamental difference between the eighteenth and the twenty-first centuries is changing our assessment and valuation of technologies, incentives and rules governing all of society's values, habits, regulations and institutions.

[34] Turner and Alexander (2014). More sources: see Jackson and Webster (2016), CC BY-NC-ND 4.0.

[35] Rockström and Klum (2012).

Economic theory therefore has to be updated, so as to adapt to the conditions of the full world. It is insufficient to incorporate environmental and social concerns by translating them into monetary expressions of capital. Nor is it sufficient to simply refer to various forms of pollution and ecosystem decline as 'externalities' – the notion being that what is at stake is some marginal disturbance. Humanity's transition into a full world also has to change the attitudes, priorities and incentive systems of all civilizations on this small planet.

Luckily, some (rare) historical evidence confirms that in mature stages of development, human happiness can improve and be maintained while the consumption of energy, water or minerals stays stable or is even reduced (see Sects. 3.1, 3.2, 3.3, 3.4, 3.5, 3.6, 3.7, 3.8, and 3.9). Economic growth and technological progress can be accompanied, if not accelerated, by an increase of elegance and efficiency of resource use, possibly in a 'cradle to cradle' manner.[36] For example, from the eighteenth-century candles to the LED, the output of light per input of energy has risen roughly a hundred million-fold,[37] allowing much more lighting convenience at much less energy consumption even in the full world.

At this moment in time, however, nearly all the trends of resource consumption, climate change, biodiversity losses and soil degradation reflect the inadequacy and misdirection of public policies, business strategies and the underlying social values. At a more basic level, these prevailing trends also reflect the inadequacy of the system of education. The cumulative implications of these trends are forcing us to *dramatically change the direction of progress and to work hard on the creation of the new Enlightenment*. That new Enlightenment should reinvigorate the spirit of inquiry and bold visioning, and a kind of humanism that is not in a primitive manner anthropocentric but allows also for compassion for other living beings, while incorporating far more attention to the long-term future (see Sect. 2.10).

Yet, this book *Come On!* is hard stuff. It will not be easy to digest. Politically, it is very uncomfortable. It both requires and represents fresh and original thought and approaches. It should be seen as an invitation to readers and discussants to 'come on', and to join on a fascinating journey of developing and testing new approaches of making the *full world* a sustainable and prospering one.

1.2 *The Limits to Growth*: How Relevant Was Its Message?

One of the main preoccupations of this book is the failure of society to understand the implications of what it means for us all to be living in a 'full world'. Therefore, it is natural to go back to 1972 and that landmark report to the Club of Rome, *The Limits to Growth* (LtG), written by Donella Meadows, Denis Meadows, Jørgen

[36] Braungart and McDonough (2002), McDonough and Braungart (2013).

[37] Tsao et al. (2010).

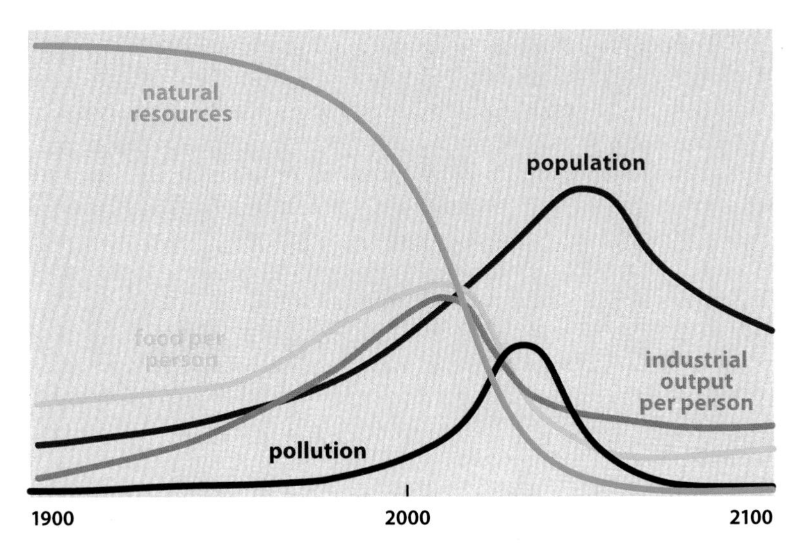

Fig. 1.2 The standard run in *The Limits to Growth*. Exhaustion of resources and heavy pollution would lead to collapse by about 2025 (Source: Meadows et al. 1972 (footnote 20))

Randers and William Behrens III.[38] The Club of Rome was, through this book, one of the first organizations to address the challenges of non-sustainable growth.

The key figure in this report was the business as usual scenario (Fig. 1.2). Assuming constant relations between natural resources, food per person, population, pollution and industrial output per person, it showed a world that would run into disaster in the first half of the twenty-first century. However, many people read the report as if the world would come to a standstill in the next 10 years or so; that was never the message. The report had established a 50- to a 100-year perspective. Moreover, its focus was the increasing physical impact of economic growth – via humanity's ecological footprint – not growth itself.

The Limits to Growth reverberated around the world and the book sold many million copies. However, massive critiques, not least by conventional economists, followed its publication. The main critique was that the report had not factored in 'the ingenuity of man'. Furthermore, economists claimed that resource scarcity is primarily a question of pricing. But critics were partially right: The treatment of innovation was too static in *The Limits to Growth*. The World3 computer model used in the MIT study was rather inflexible and assumed constant mutual relations between different parameters such as industrial output and pollution.

The model could not predict the stunning advances in pollution control, which permitted many countries to partially escape from tragedies of polluted air, water and soils. This being said, there are of course limits to what technology can achieve.[39]

---

[38] Meadows et al. (1972).

[39] Higgs (2014, pp. 51–62; 257–268).

With regard to resource scarcity the picture is mixed. Renewable resources tend to be overexploited, like overfishing, groundwater depletion or deforestation, as well as ecosystem degradation and pollution. For non-renewable resources the picture is more complex. Some materials, like iron ore, remain abundant. For others, like certain metals and phosphorus, there is no doubt a risk of scarcity. A common problem is that once the richest ores are exploited, further extraction will require increasingly more energy and generate more pollutants.[40]

Despite some shortcomings in the World3 computer model used, it was never correct for conventional economists to dismiss the warnings of the report. Their understanding of the functioning of the natural world was – and still is – limited. They seem to make no distinction between financial and industrial capital on the one hand and natural capital on the other. These types of capital are treated as near perfect substitutes for one another. 'As long as financial capital increases we are fine' – so goes the thinking. But we cannot eat money and money cannot generate more orangutans or clean water or a stable climate, once overuse or pollution has gone too far.

Furthermore, conventional economic models, linear in nature, are incapable of addressing and guiding society with regard to the non-linearity of natural systems, such as the climate system. Scientists keep reminding us of 'tipping points' in relation both to vital ecosystems like rainforests, soils or lakes and the climate system. Once such tipping points are crossed and the original ecosystem has flipped or the climate system is severely destabilized, the damage made may be irretrievable. Examples include hydrocarbon leakage from the melting tundra in Siberia, the bleaching of coral reefs and parts of the Amazonian rainforest tipping over and becoming a savannah.

Shortly after the publication of *Limits*, the oil exporting countries (OPEC) boldly made use of their near-monopoly in oil and gas, and through concerted action managed to quadruple oil prices. This oil shock, however, triggered intensified search for more oil resources, and less than 10 years later supplies exceeded demand, so oil prices began to tumble again. Environmental optimists and especially conventional economists saw this as proof of their critique of *Limits*. During all of the 1980s and 1990s, the Club of Rome's message of *The Limits to Growth* enjoyed very little mainstream appreciation and attention.[41]

Nevertheless, the core of the message remained valid. When the new industrial giants, China and India, entered the world commodity markets in a massive way, demanding increasing amounts of fossil fuels, cement and metallic minerals, the prices of those commodities began to rise again, and a new era of scarcity seemed to have begun. However, in the wake of the economic crisis of 2008, prices collapsed yet again (Fig. 1.3).

---

[40] Bardi (2014).

[41] Higgs (2014, l.c., p. 91–93).

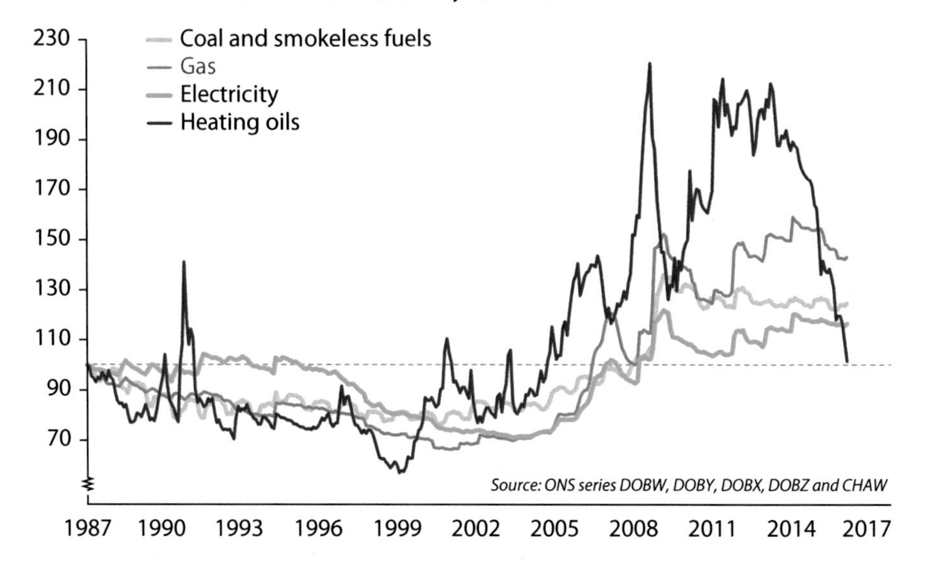

Fig. 1.3 Index energy prices rose from 2004 until the end of 2008, and again till 2014 but collapsed again later (Source: Dempsey et al. 2016)

A recent study by Graham Turner found that historical data from the period 1970–2000 again confirmed the predictive value of *Limits*.[42] While most decision makers have tended to dismiss that message in favour of more optimistic scenarios, being politically popular, it remains our conviction that *The Limits to Growth* in essence is right, after all.

1.3 Planetary Boundaries

The idea of planetary boundaries has proven a very effective means of gauging the state of the planet. This concept was introduced in 2009 by a group of 28 internationally renowned scientists led by Johan Rockström and Will Steffen and has been recently updated.[43] The concept indicates, based on scientific research, that since the Industrial Revolution human activity has gradually become the main driver of global environmental change. Once human activity passes certain thresholds or tipping points (defined as 'planetary boundaries'), there is a risk of 'irreversible and abrupt

[42] Turner (2008–09).

[43] Rockström et al. (2009a, b). See also Steffen et al. (2015).

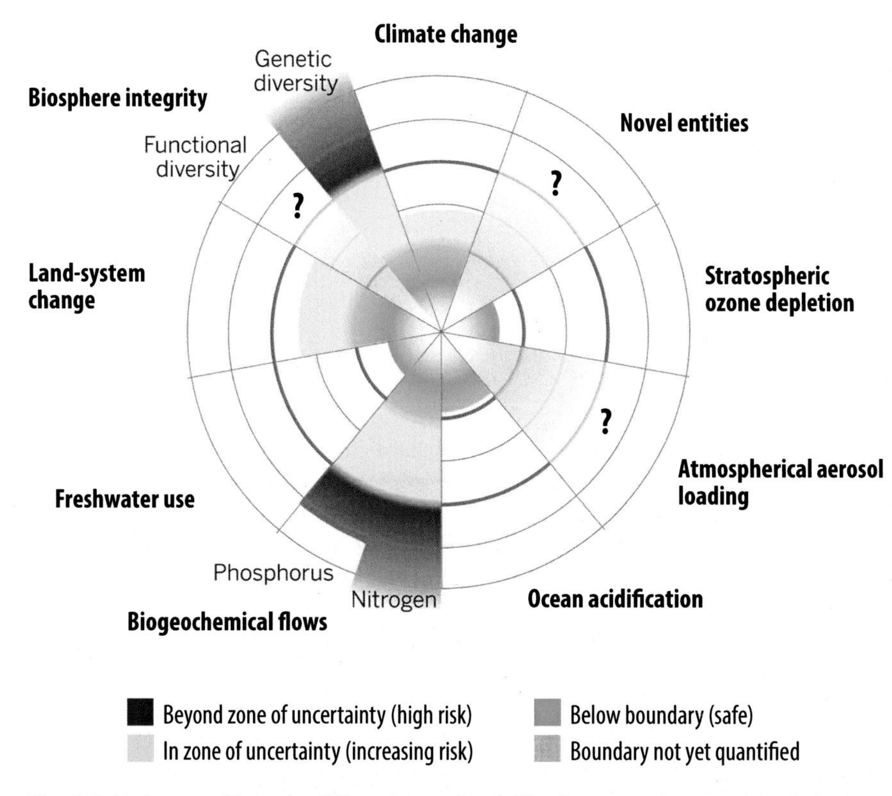

Fig. 1.4 Estimates of how the different control variables for seven planetary boundaries have changed from 1950 to present. The green-shaded polygon represents the safe operating space (Source: Steffen et al. 2015; http://science.sciencemag.org/content/347/6223/1259855)

environmental change'. Rockström et al. identified nine 'planetary life support systems' essential for human survival and attempted to quantify how far they have been pushed already.

The nine planetary boundaries are shown in Fig. 1.4 and the ensuing list.

The list is a bit clearer:

- Stratospheric ozone depletion
- Loss of biodiversity and extinctions
- Chemical pollution and the release of novel entities
- Climate change
- Ocean acidification
- Land system change
- Freshwater consumption and the global hydrological cycle
- Nitrogen and phosphorus flows to the biosphere and oceans
- Atmospheric aerosol loading

Without commenting on details of all planetary boundaries, this book will address the most prominent issue, climate change, in Sect. 1.5.

1.4 The Anthropocene

One of the most striking ways of describing the current, human-dominated era is the calculation that humans and farm animals (Fig. 1.5), combined, constitute 97% of the bodyweight of all living land vertebrates on earth! This means elephants and kangaroos, bats and rats, birds, reptiles and amphibians combined make up a mere 3% of the world's land vertebrate bodyweights.

The source of this stunning observation is from critics of excessive meat eating.[44]

It is pretty clear that humankind's steeply rising consumption rates, especially during the past 50 years, has caused massive changes to the atmosphere and the biosphere. The effects on human health are yet to be quantified, although there is abundant anecdotal evidence of quite deleterious effects.

Atmospheric chemist and Nobel Prize winner Paul Crutzen says a more scientific way of appreciating that the Anthropocene era is now underway is to look at the curves describing the changes of a variety of parameters, both physical and social, observed over the past 250 years. Figure 1.6 shows how 24 such parameters developed, with the growth drama occurring during the past 50 years.[45]

Fig. 1.5 Factory farming (pigs in this picture) is the main reason for the fact that 97% of the living vertebrate biomass on land is farm animals and humans. Three per cent remains for wildlife (Source: Getty Images/iStockphoto/agnormark)

[44] World Society for the Protection of Animals (WSPA) (2008). Figures are based on Smil (2011).

[45] Steffen et al. (2007).

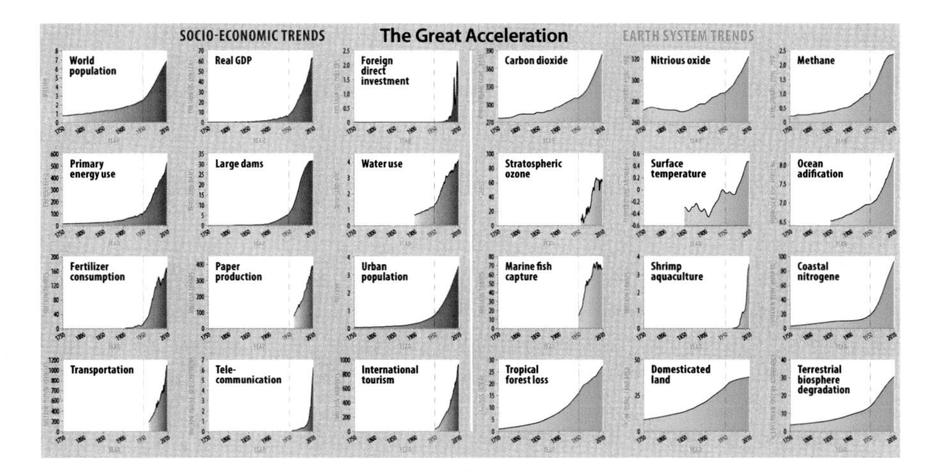

Fig. 1.6 The Anthropocene. Twenty-four curves showing the dramatic changes of human population, of the chemical composition of the atmosphere and of human construction and consumption patterns. The dramatic changes occurred during the past 50 years (Adapted from Steffen et al. 2007; by courtesy of Globaïa, www.globaia.org)

It does not take an unusual imagination to conclude that such massive changes have the potential of leading to violent conflicts, probably on a scale not seen any time in the past. Clearly, under war conditions, achieving any one of the 11 socio-economic SDGs (Sect. 1.10) would become impossible. Thus, for the sake of the socio-economic well-being of humankind, it is absolutely imperative that the world avoid the kind of environmental disasters resulting from trespassing the planetary boundaries.

1.5 The Climate Challenge

The 21th Conference of the Parties (COP 21) of the UN Climate Convention in Paris in December 2015 was hailed as a big success. All 195 countries present in Paris did agree on the need 'for global emissions to peak as soon as possible' and to 'undertake rapid reductions thereafter'. The call to hold the increase in global average temperature 'well below 2°C and to pursue efforts to limit the increase to 1.5°C above pre-industrial levels' is no doubt very ambitious.

For all the official praise, there were also quite a lot of critical comments. Leading climate scientist Jim Hansen called the agreement a fraud. 'It's just worthless words. There is no action, just promises. … As long as fossil fuels appear to be the cheapest fuels out there, they will be continued to be burned. … The decision is meaningless without a commitment to tax greenhouse emissions', he said to *The Guardian*[46]; Hansen believes that a strong price signal is the only way to reduce emissions fast enough.

---

[46] The Guardian. 13 Dec., 2015.

George Monbiotl summarized it otherwise, also in *The Guardian*: 'The deal is a miracle by comparison to what it could have been – and a disaster by comparison to what it should have been'. He added, 'The real outcomes are likely to commit us to levels of climate breakdown that will be dangerous to all and lethal to some'.[47]

Monbiot's comments must be taken seriously. It was, indeed, an achievement to agree, not only to keep temperature increase 'well below 2°C' but also to aim 'to limit the increase to 1.5°C'. However, hardly anything is said about what measures to take to achieve these goals. No agreement was reached on the necessity of a global carbon tax, nor on the phasing out of fossil fuel subsidies. Furthermore, the pace foreseen in terms of emission reductions in the years leading up to 2030 – a critical period to avoid accumulating excessive amounts of CO_2 in the atmosphere – is modest, at best. There does seem to be a serious disconnect between what is being done and planned for and what is required.

If countries mainly stick to their Paris commitments – the so-called INDCs (Intended Nationally Determined Contributions) – there is little chance of preventing the average global temperature of reaching a minimum of 3 °C above pre-industrial levels, as early as the second half of this century. Such warming could be catastrophic. The climate system is non-linear in nature and may reach unpleasant tipping points already around a warming of 1.5 °C or 2 °C. This makes action in the immediate future so important.

1.5.1 We Need a 'Crash Plan'

Let's face it. To have a chance to meet the Paris goals, the world has to go through a rapid and thorough transformation of its production and consumption systems. To avoid exceeding the 2 °C target, the carbon intensity of the global economy must be reduced by at least 6.2% per annum. To meet the 1.5 °C target the reduction would have to be close to 10% yearly. To put this in perspective, global carbon intensity fell by an average of 0.9% between 2000 and 2013!

A positive sign is that many smaller but still key actors – states, cities, companies, financial institutions, civil society organizations, faiths and communities – have lined up in support of the Paris agreement. More than 1000 cities around the world are committed to 100% renewable power, and the same goes for almost 100 of the world's largest companies.

But the challenge is colossal – not least in an open and market-based economy. Humanity truly needs a 'crash plan'. One thing seems obvious: the market alone will not solve the problem. Averting climate change will require such large-scale, rapid action that no single technology, new or emerging, can be the solution. The challenge therefore is one of rapid, concerted deployment of a portfolio of emerging and mature energy and non-energy technologies. For this to happen, governments – not short-term-focused markets – must be in the driver's seat.

[47] Monbiot (2015).

It could be argued that society has the knowledge, the financial resources and the technologies to move towards a low-carbon society in time to avert disaster. With learning curves for solar and wind – and more recently for energy storage – being extraordinarily positive, there is no longer any excuse for not taking strong action.

But lower technology costs alone will not make it. All the sunk costs in power plants, vehicles and manufacturing facilities designed to run on fossil fuels are effective barriers to change. The incumbents, no doubt, will do all they can to prevent or at least postpone the necessary transition. The absence so far of a global tax on carbon and the price of oil hovering around US$50 per barrel will not make change any easier.

Few people want to talk about it. But the truth of the matter is that if humankind does not manage to put in place the 'crash plan' needed for the decarbonization of the economy, there are two alternatives left, both highly questionable in terms of efficacy and with unknown ecosystem effects: geoengineering and the large-scale deployment of 'negative emissions technologies'.

1.5.2 How to Deal with Overshoot?

Carbon dioxide is long-lived in the atmosphere and the remaining carbon budget is extremely tight. It is therefore realistic to assume that CO_2 emissions will overshoot. The question is by how much?

The Paris agreement pledges to arrive at greenhouse gas neutrality by 2050. The wording can be read as an invitation to 'geoengineering', from the comparatively harmless but expensive CCS (carbon capture and sequestration) and biogenic CCS (BECCS) to wild phantasies of manipulating the atmosphere, the stratosphere or the ocean surfaces with a view of changing the global radiation patterns so as to reduce average temperatures.

Within the Club of Rome, there are strong voices in favour of CCS, arguing it is the only method that has a chance of stopping run-away climate change. On the other hand, both for technical CCS and for BECCS the scale needed to make a difference is enormous. The following comment by Professor Kevin Andersson, guest professor at Uppsala University and deputy director of the Tyndall Centre, puts BECCS into perspective:

> The sheer scale of the BECCS assumption underpinning the Paris Agreement is breathtaking – decades of ongoing planting and harvesting of energy crops over an area the size of one to three times that of India. At the same time the aviation industry anticipates fuelling its planes with bio-fuel, the shipping industry is seriously considering biomass to power its ships and the chemical sector sees biomass as a potential feedstock. And then there are 9 billion or so human mouths to feed. Surely this critical assumption deserved serious attention within the Agreement.[48]

[48] Kevin Andersson. 2015. The hidden agenda: how veiled techno-utopias shore up the Paris Agreement. Pre-edited version of his summary of the Paris Agreement published in Nature's World View (Dec. 2015): http://www.nature.com/polopoly_fs/1.19074!/menu/main/topColumns/topLeftColumn/pdf/528437a.pdf

Add to that the logistical, legal and public acceptance question marks. The volumes of CO_2 to be stored to compensate for carbon overshoot are extraordinarily large in most of the IPCC 2 °C pathways. Unfortunately, limited efforts have been devoted to critically analysing whether such volumes are at all possible to obtain. No doubt, strong efforts must be made to develop CCS technology further, as it will be needed as a fall-back strategy to address carbon emissions. It is not possible to ignore the continued use of coal in many parts of the world in the foreseeable future, as well as the production of steel and cement.

1.5.3 Why Not a Marshall Plan?

It is very true that negative emissions will also be needed, and BECCS is an option here. But everything must be done to limit its scope because a huge reliance on 'negative emissions technologies' is dangerous. It tends to give people a false sense of security that society will find a way to engineer a solution for the climate problem.

Instead of agreeing on something like a Marshall Plan to invest massively in low-carbon technologies, which is possible from both a technological and an economic point of view, the Paris agreement assumes that mitigation measures in the period leading up to 2030 would only deliver reductions in the range of 2% p a. If climate change is a serious threat – and the Paris agreement says it is – prudence would compel us to take much stronger action in the immediate future and not leave it for later. Without such action, the reliance on negative emissions would be dangerously high.

The main hope for the post-Paris agenda is that different actors (governments, cities, companies, financial markets and civil society organizations) will take the challenge seriously and do everything possible, right now to help support a strengthening of mitigation efforts across the board. Strong action by individual governments, states or cities does matter. The world is in desperate need of good examples, including in your own neighbourhood.

1.5.4 Has Humanity Already Missed the Chance to Meet the Climate Goals?

Almost 2 years have passed since Paris. The year 2016 alone brought a great number of stories related to human-caused climate change – some good, some bad and some downright ugly.

On the positive side, there is the fact that the Paris agreement was ratified much faster than most have believed. The parties to the climate convention met again in November 2016 in Marrakech. Many observers feared that a number of governments would use Trump's victory (which happened during the conference) as a pretext for reducing their ambitions in terms of emission reductions. On the contrary, major governments, including the United States (with President Obama still in charge) and China, reiterated their commitments from COP 21 and urged the world community to enhance their efforts to meet the Paris goals.

Furthermore, at a meeting in Kigali, Rwanda, about a month before Marrakech, in October 2016, nearly 200 countries struck a landmark deal to reduce the emissions of one of the most powerful greenhouse gases, hydrofluorocarbons (HFCs), a move that could prevent up to 0.5 °C of global warming by the end of this century.

Probably the best news of all is the rapid cost reductions and expansion of clean energy – mainly solar and wind – all over the world. 'World energy hits a turning point' was a Bloomberg headline.[49] 'Solar power, for the first time, is becoming the cheapest form of new electricity', the article marvelled (see Sect. 3.4).

On the negative side, and in spite of the progress referred to above, global warming continues. The year 2016 crushed the record for the hottest year, set back in 2015, which itself smashed the previous record, set in 2014. Joe Romm of Climate Progress comments, 'Such a three-year run has never been seen in the 136 years of temperature records. It is but the latest in an avalanche of evidence in 2016 that global warming will be either as bad as climate scientists have been warning for decades – or much worse'.[50]

If temperature records fail to convince people about the trend of warming, several studies in 2016 brought new evidence of the extent to which the oceans are warming. The excess energy stored in the oceans is huge and means that much of the energy surplus on the earth is here to stay for centuries.

The year 2016 was a crazy year in terms of climate-change-induced weather events. There were severe droughts in many parts of the world and serious flooding in others. There was an incredible heat wave in the Arctic, which led to the lowest wintertime ice ever recorded. Hurricanes and typhoons grow stronger with global warming. According to expert Jeff Masters,[51] the strongest storms ever measured occurred in two regions in 2016, along with seven Category 5 storms, a huge number for a single year. The trend has continues in 2017 with massive tropical storms in Asia and the Americas – *Harvey* and *Irma* causing huge devastation in Texas and Florida.

When it comes to the truly ugly events, it should come as no surprise that the election of Trump was the most important. Some observers hoped that President Trump would eventually start listening to the scientists and take climate change seriously. However, his decisions in favour of coal, oil and gas in March 2017 don't support such optimistic hopes. Even worse, of course, was his decision in early June to withdraw the United States from the Paris agreement.

Climate change is an issue where international agreements are indispensable. It took the world 23 years from the Earth Summit in 1992 – and the signing of the climate convention – to reach such an agreement. The United States – under Obama – played an important role in making the agreement possible. Trump's decision is no less than a tragedy for the climate convention and the efforts made by many governments, cities, companies and civil society organizations across the

---

[49] Bloomberg New Energy Finance. 2016.12.15. World Energy Hits a Turning Point.

[50] Romm (2017).

[51] Jeff Masters. 2016. The 360 Degree rainbow Jeff Masters Blog December 2016.

world to prevent dangerous climate change. His behaviour is both arrogant and ignorant. While other governments have agreed to put the climate first, he insists on putting America first. The irony is that the United States no doubt will come out as a loser, both with regard to its position in world politics – abandoning its leadership role – and its lead position in the development of clean technology. Other countries – not least China – will take over.

As already noted, the pace of emission reductions in the coming years must go far beyond what was foreseen in the Paris agreement. If not, there is no chance whatsoever to meet the Paris objectives. Without active participation from the United States the challenge will be colossal.

In conclusion, our view on the Paris agreement and the possibilities of keeping the global temperature increase 'well below 2 °C' is considerably more pessimistic today than it was a year ago. The election of Trump – and his actions aimed at prolonging the fossil-based economy and enriching the owners of fossil energy – is one important factor; the other is that so far, very few governments have stepped up to the challenge of the ambitious goals that were set in Paris and have reconsidered their INDCs. The world is still on a pathway to at least 3 °C of warming.

To have a chance to save the Paris agreement and prevent dangerous climate change, players like the European Union, China and India must from now on assume a much more proactive role in climate policymaking. The EU did provide leadership on climate for the last two decades, not least during the presidency of George W. Bush. Now the world is in a similar, if not worse, situation.

For the EU to take on the leadership role again, the targets set by the EU for 2030 – a reduction of GHG emissions by 40% compared to 1990 – are totally inadequate. Even China and India have to revisit their goals and develop more ambitious targets. Parallel to that, action must be considered with regard to what measures to take in terms of border tariffs to offset the advantage products produced in the United States will have compared to regions whose companies are exposed to carbon taxes or emissions trading. We shall revert to these challenges, both with regard to the SDGs and the Paris agreement, in Chap. 3 of this book.

1.6 Other Disasters Ahead

1.6.1 Technological Wildcards and Familiar Threats

The Cambridge (UK)-based Centre for the Study of Existential Risks (CSER), founded in 2012, has quickly gained worldwide visibility through presenting a number of dangers that could even lead to the extinction of humanity. Of course, these can include astronomical catastrophes, such as the earth's collision with a giant meteorite, or the appearance of a deadly and extremely contagious microorganism against which no remedy can quickly be found. But realistically, the group, led by Seán Ó hÉigeartaigh, also investigates technological development entirely designed by humans. Ó hÉigeartaigh calls them technological wildcards.[52] They include the following:

- Synthetic biology creating viral and bacterial organisms with novel and deadly characteristics and capabilities that could infect humans and spread around the world; a particularly controversial area of research is 'gain-of-function' virology research, leading to viruses with completely unknown capacities. More conventional is the unintended spread of multi-resistant microorganisms caused by the preventive overuse on farm animals of antibiotics, or by high antibiotic concentrations in poorly treated wastewater from antibiotic drug industries.[53]
- Geoengineering: a suite of proposed large-scale technological interventions that would aim to 'engineer' our climate in an effort to slow or even reverse the most severe impacts of climate change. Seemingly, President Trump intends to spend a lot of money on geoengineering.[54]
- Advances in artificial intelligence capable of matching or surpassing human intellectual abilities across a broad range of domains and challenges (see Sect. 1.11.3).

Obviously, humanity must respond to such absolutely scary prospects. Technology assessment is the very least that must be done. Prohibitions against research that could lead to the extinction of the human race must also be considered (see Sect. 3.15.2).

Quite different are potential disasters that in a sense are familiar. A web search[55] for 'economic collapse' will return almost 35 million sources of information. Depressing literature is widely available. The challenges are not confined to serious disturbances to the atmosphere and the biosphere. Major *social* challenges were already addressed in Sect. 1.1.

Early in 2016 the principal geologist of the British Geological Survey stated that human-caused changes to the earth are greater than the changes that marked the end of the last ice age.[56] A problematic chemical, perfluorooctanoic acid, is now found in the tissues of polar bears and all humans on earth. Plastics are found in the guts of 90% of seabirds,[57] and microparticles, the decomposition of the millions of tons of plastic waste generated every year, are now ubiquitous.[58] Ninety per cent of all the oil consumed by humans has happened since 1958, and 50% of it since 1984[59]; and that has left a permanent record of black carbon in glacial ice.

[52] Sean Ó hÉigeartaigh (2017).

[53] Lübbert et al. (2017).

[54] https://www.theguardian.com/environment/true-north/2017/mar/27/trump-presidency-opens-door-to-planet-hacking-geoengineer-experiments

[55] Economic Collapse, Google, accessed September 2016.

[56] Waters et al. (2016).

[57] The Guardian, http://www.theguardian.com/environment/2015/sep/01/up-to-90-of-seabirds-have-plastic-in-their-guts-study-finds Associated Press, 1 Sept 2015,

[58] Hasselverger (2014).

[59] BP Statistical Review of World Energy 2006.

In a rather extreme projection, Walter and Weitzman[60] describe the economic shocks that are likely to result from climate change. They expect massive disruptions for agriculture and consequently for nutrition, potentially destroying much of the hopes contained in the SDG 2 (see Sect. 1.10).

Much less concretely described but potentially equally disastrous are the massive losses of biodiversity. Already today, the earth is in the midst of the '6th extinction event'.[61] The first five were caused by tectonic and volcanic events on a geological time scale; in the case of the dinosaurs, an astronomical catastrophe is also thought to have played a key role. But the sixth, unfolding very rapidly over the last century, is exclusively caused by humans. During this period, an explosive increase in human population plus an ever-increasing land use (see the 'footprints' saga, Sect. 1.10) have destroyed or completely altered most habitats of wild plant and animal species. Small wonder that some hundred animal and plant species are lost every day, the majority of which have not even been scientifically identified before their extinction. The effects on humans of this tragedy will most probably be very dangerous, but details are hard to predict. In his latest book, E.O. Wilson suggests that half of the earth's surface should be reserved for nature's protection[62] – not exactly realistic under the conditions of further human population growth.

Soil erosion, soil degradation, droughts, floods and invasive species can massively add to the dangers confronting future generations. Industrialized agriculture using 'systemic pesticides' such as neonicotinoids is a deadly threat to honeybees and other pollinators.[63] There is also increasing evidence about pesticide residues in various food products. The question cannot be avoided: For how long can biological systems be managed the same way as industrial production? The long-term effects on soils by decades of pesticides being spread on them is a major issue and has so far been poorly researched. If both bacteria and fungi are lost, the soil degrades. 'Every time the soil is disturbed, or artificial fertilisers and pesticides are applied, soil life is killed and soil structure compromised', says soil scientist Elaine Ingham.[64]

Another disturbing and complex issue has to do with the production of biofuels. When biofuels are produced from residue materials from forestry and agricultural production, the benefits are clear. However, when fertile soils, like in the United States, or in virgin forests, as in Indonesia, are turned into large-scale monocultures of maize or palm oil, the negative social or environmental consequences can far outweigh the positive ones.

Another new and disquieting technical concern is human-engineered 'gene drives'.[65] A successful gene drive is capable of intentionally or accidentally altering a species or causing its extinction. So far, these artificial gene drives are developed

[60] Walter and Weitzman (2015).

[61] Kolbert (2014).

[62] E.O. Wilson 2016. Half Earth: Our Planet's Fight for Life.

[63] E.g. van der Sluijs et al. (2015).

[64] Elaine Ingham. 2015. The Roots of Your Profits (video).

[65] National Academies of Sciences, Engineering, and Medicine (2016).

using the new 'gene-editing' system known as CRISPR-Cas9. Gene drives may be deliberately introduced into invasive species to eradicate them from the wild for conservation purposes, or into weed species to remove them from farmers' fields: all desirable plans at first glance. But gene drives could just as easily be pressed into use for military purposes, as bioweapons, or to suppress food harvests. There are unintended effects as well; 'because gene-drive modified organisms are intended to spread in the environment, there is a widespread sense among researchers and commentators that they may have harmful effects for other species or ecosystems'.[66] There is no internationally agreed process for the effective governance of trans-boundary effects arising from the release of a gene drive, an enormous governance gap. In consequence, more than 160 NGOs mostly from developing countries, present in Cancún at the 13th Convention of the Parties of the UNCBD (Convention on Biological Diversity) in December 2016, demanded a moratorium on applied research, development and release of genetically engineered gene drives.[67]

Then there are the political dangers, touched upon in Sect. 1.1. Wars and conflicts rage in the Near East, in some African countries and in Afghanistan and Myanmar. They have led to unprecedented migrations of refugees, both inside and outside the actual war-torn regions.

Political disasters are often linked with nature. Climate change is a partial cause for conflicts over water and fertile soils. And then, let us not ignore the fact that wars tend to happen in the regions with the highest population growth. Of course, that was also the case in the 'empty world', but in the 'full world' there is no easy way out, thus enhancing the conflicts on resources. Moreover, in earlier times, even the poor were embedded in a basically benign, robust and fertile planet. That is no longer the case.

1.6.2 Nuclear Weapons: The Forgotten Threat[68]

A nearly forgotten threat is the spectre of nuclear weapons. Nuclear weapons are the most deadly of all mass killing devices. They put civilization, the human future and the future of life on the planet at serious risk. They are illegal, immoral and waste resources that otherwise could be used to meet human needs. Humankind needs to find a path to abolish nuclear weapons before these weapons abolish us.

And yet, since the end of the Cold War, nuclear weapons have generally been viewed with complacency by the world's societies. These weapons, in the arsenals of nine countries, are largely kept out of sight and out of mind. To the extent that possessing and threatening to use nuclear force makes it into the public consciousness

---

[66] Ibid. doi:10.17226/23405

[67] Civil Society Working Group on Gene Drives (2016).

[68] The alert to this deadly danger came from the Nuclear Age Peace Foundation and its President, Dr. David Krieger, Member of the Club of Rome. See their homepage https://www.wagingpeace.org/.

and discourse, they are justified on the grounds of nuclear deterrence, that is, the threat of nuclear retaliation. But that remains an unproven hypothesis about human behaviour and a potentially destabilizing one at that.

Society is beginning to forget that an all-out nuclear war could lead to a *Nuclear Winter*, potentially sending temperatures to their lowest levels in 18,000 years, triggering an ice age, and destroying a large part of life on earth.

The Nuclear Non-Proliferation Treaty (NPT) of 1970 divided the world into nuclear 'haves' and 'have-nots'. As defined by the NPT, the nuclear (NPT) are those countries which had manufactured and exploded a nuclear weapon prior to January 1, 1967. France and China were added to the nuclear 'haves', when they later joined the treaty. Three countries never joined the treaty – Israel, India and Pakistan – and went on to develop nuclear arsenals; and one country, North Korea, withdrew from the treaty in 2003 and is playing an evil poker game building up a nuclear arms arsenal.

All nine nuclear-armed countries are now engaged in modernizing their nuclear arsenals. The United States plans to spend $1 trillion doing so over the next three decades. Other nuclear-armed states also have ambitious modernization plans. The waste of resources and lost-opportunity costs are staggering. Beyond this, however, modernization of nuclear arsenals is making the weapons smaller, more accurate and more efficient. All this sums to making the weapons more usable by military commanders and thus more likely to be used. Modernizing nuclear arsenals is a clear violation of the NPT (Fig. 1.7).

Jonathan Granoff of the Global Security Institute adds: If less than 1% of the 14,000 nuclear weapons in the arsenals of the nine possessor states in the world were to explode, tons of debris would enter the stratosphere, lower the earth's temperature, destroy the stability of the ozone layer, cause cancers and other horrible diseases to spread, and end agriculture as we know it. In sum, a nuclear exchange of

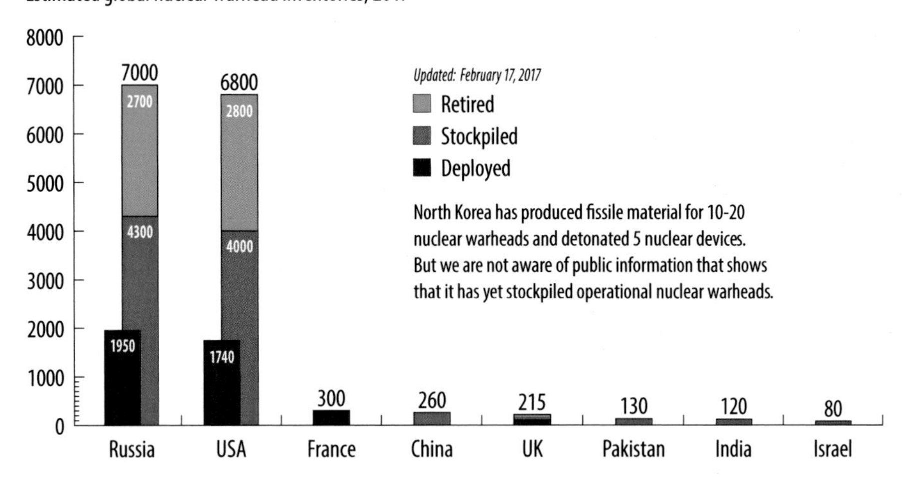

Fig. 1.7 The world's nuclear arsenals, 2017 (Source: Kristensen and Norris 2017; https://fas.org/issues/nuclear-weapons/status-world-nuclear-forces)

the arsenals of only two of the nuclear powers, say India and Pakistan, could end civilization everywhere – as would a robust first strike from the arsenals of Russia or the United States.[69]

A quarter century after the end of the Cold War, some 2000 nuclear weapons remain on high alert, ready to be fired within minutes of an order to do so, meaning that civilization could be destroyed in a single afternoon of nuclear exchange. In July 2016, an International Peoples' Tribunal on Nuclear Weapons and the Destruction of Human Civilisation was held in Sydney, Australia, condemning politicians and the nuclear weapons industry for violating human rights by still 'modernizing' nuclear arsenals and seriously considering the use of these weapons.

The threat is global and the solution must also be global. It will require negotiations with the aim of truly prohibiting and eliminating nuclear weapons. These will not be easy, as there will be many interests at the bargaining table. It will require a new legal instrument for the phased, verifiable, irreversible elimination of nuclear weapons. It must result in a treaty that accomplishes the elimination of nuclear weapons, without leaving the world dominated by conventional forces. In the end, it must be a treaty that changes the dynamics of the planet from the insanity of Mutual Assured Destruction (MAD) to the needed new reality of Planetary Assured Security and Survival (PASS).

1.7 Unsustainable Population Growth and Urbanization

Figure 1.13 in Sect. 1.10 has two horizontal dotted lines. The upper one is the 'world biocapacity of 1961', that is, the allowed ecological footprint per capita in a world populated by 3.1 billion people. The lower line is the biocapacity 2012, with a population of 7 billion people. The situation would be hugely more comfortable if the world population had stabilized 50 years ago at below 3.5 billion. However, most demographers believe that stabilization will not occur before the second half of our century, and then at a number rather beyond 10 billion. When addressing sustainable development, it is simply unavoidable to consider the question of world population. This is extremely delicate politically.

1.7.1 Population Dynamics

The old industrialized countries had their steep population increases during the nineteenth century and solved their domestic problems of overpopulation by conquering other parts of the world, notably in the Americas, Africa and Australia, and letting large numbers of people emigrate there. Thus for them to admonish developing countries to stop growing is a political non-starter.

---

[69] Widely circulated email by granoff@gsinstitute.org, dated 16 Dec., 2016.

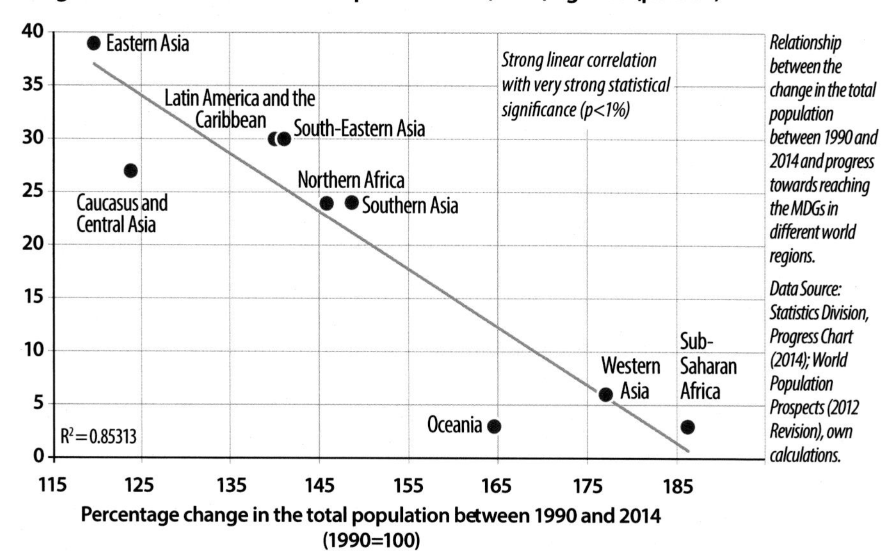

Fig. 1.8 Strong population growth is correlated with weak development (Source: Michael Herrmann (editor). 2015. Consequential Omissions. How demography shapes development – Lessons from the MDGs for the SDGs. New York and Berlin: United Nations Population Fund and the Berlin Institute for Population and Development, UNFPA 2015)

However, it is both legitimate and fruitful for developing countries themselves to think of ways and means of arriving at a sustainable population policy.

The United Nations Population Fund (UNFPA) has published a new study[70] confirming the positive correlation between economic success and restraint on population growth (Fig. 1.8). Regions with rapid population growth are associated with weak development, although of course the logic of this correlation can work both ways. Nevertheless, it is an established fact that in most cultures, reaching a high level of development, that is, enjoying adequate education, employment and self-determination of women, as well as having access to plentiful energy, leads to the stabilization of that group's population. Conversely, policy makers and religious leaders must realize that strong population growth tends to weaken the economic development of their countries.

On a finite planet, population growth should be curtailed before nature forces the issue. The Club of Rome commends those countries which have sought rapidly reduced reproduction rates, and congratulates them for having actively promoted the programmes that are proven to achieve this, that is, health care for infants and children

[70] UNFPA. 2015. Consequential omissions. How demography shapes development – Lessons from the MDGs for the SDGs. Fig. 8; electronic source: http://www.berlin-institut.org/fileadmin/user_upload/Consequential_Omissions/UNFPA_online.pdf

Population projections

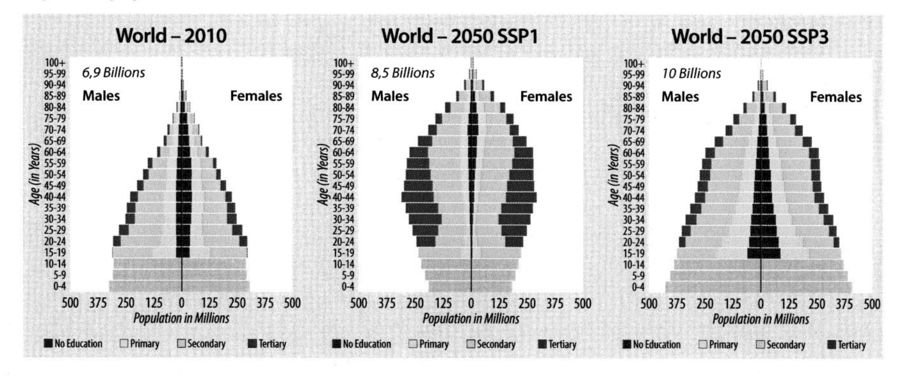

Fig. 1.9 Different population projections in 2050, depending on the education profile of the population. The middle projection 'SSP1' refers to a high education scenario ending up with 8.5 billion people in 2050, while the 'SSP3' projection with low education levels ends up with 10 billion people in 2050 (After: KC S, Lutz W (2014) Demographic scenarios by age, sex and education corresponding to the SSP narratives. Population and Environment 35 (3): pp. 243–260. DOI: 10.1007/s11111-014-0205-4)

under five, reproductive health services, including family planning, women's education and emancipation, as well as working towards increased per capita prosperity and providing some social security for the elderly, all of which help eliminate some of the impetus for having large families.

A recent study of KC and Lutz[71] estimates that better education can lead to one billion fewer people in 2050 than currently anticipated (see Fig. 1.9). Many developing countries have committed themselves to the empowerment of women through education and economic inclusion as part of their quest for sustainable development. It is imperative for development cooperation to focus on achieving the desired outcomes in this area.

Wealthier countries had committed themselves in the 1994 Cairo Plan of Action to provide reproductive health services and family planning, but neither national governments nor donors have so far lived up to these Cairo promises. This means that an estimated half million women, worldwide, still die during childbirth each year. Hundreds of millions of couples lack access to contraception, a situation that the Catholic Church, until recently, helped to cement. Although many more children are attending school today than 10 years ago, a large gap still remains between boys and girls. In countries like India, Nepal, Togo, Yemen and parts of Turkey, there are 20% more boys than girls in school. In the poor rural districts in Pakistan, the proportion of girls being educated is less than a quarter.

In many developing countries, the number of births per woman is still between four and eight. The primary cause is poverty, but the low status of women in society also plays a major role, and all forms of discrimination against women remain a huge

[71] Guttmacher Institute. (authors: Jacqueline E. Darroch, Vanessa Woog, Akinrinola Bankole and Lori S. Ashford) 2016. Adding It Up: Costs and Benefits of Meeting the Contraceptive Needs of Adolescents.

problem. India has launched TalentNomics to measure the economic costs/benefits of gender gaps, with a view of boosting women's opportunities.[72]

Regarding environmental impacts linked to population growth, it is apparent that human numbers per se do not tell the whole story. The 'I = PAT equation' worked out by Paul Ehrlich and John Holdren[73] names three factors affecting human environmental impacts (I): population numbers (P), relative affluence (A) and technology use (T), with T representing the hope of dramatically reducing environmental impacts per unit of added value (see Sects. 3.4, 3.8 and 3.9).

The recent age of the 'great acceleration' (see Fig. 1.6) clearly demonstrates that population alone does not explain the massive increase of human impact: while human numbers grew only fivefold, world economic turnover grew 40 times, and fossil fuel use 16-fold. Fish catches grew by a factor of 35, and human water use 9-fold.

While population numbers are but one of the factors explaining the growing footprints of mankind, it is crucial to increase the efforts worldwide – and not least in Africa – to encourage families to reduce the number of births. The challenge of addressing climate change as well as ecosystem decline will be considerably more doable if the world population levels out around 9 billion – which would still be possible – than between 10 and 11 billion or beyond.

1.7.2 Urbanization

Humanity is turning from a rural to an urban species. Global urbanization is seemingly unstoppable worldwide (Fig. 1.10). In developed and developing countries, cities offer easier access to resources and job opportunities than rural areas, as well as cultural, education and health benefits. As centres of economic power and social interaction, and of both production and consumption, they have a magnetic attraction.

In 1800, there was just one city of a million people – London. From that time onwards, global urbanization, closely linked to Industrial Revolution technologies, got under way. From 1900 to 2011, the global human population increased 4.5-fold, from 1.5 to 7 billion. During that time the global urban population expanded 16-fold, from 225 million to 3.6 billion, or to about 52% of the world population. By 2030, 60% of the world population, or 4.9 billion people, are expected to live in urban areas, over three times more than the world's entire population in 1900.[74]

Today, there are more than 300 cities of one million people or more and 22 megacities of over 10 million people, with 16 of them in developing countries.[75]

[72] See IMF's Gender Income Gap Studies, Japan and Mckinsey Gender Gap Income Loss Analysis all done in 2015; see Google for details.

[73] Ehrlich and Holdren (1971).

[74] United Nations (2011).

[75] World Resources Institute Washington, Urban Growth, www.wri.org/wr-98-99/citygrow.htm

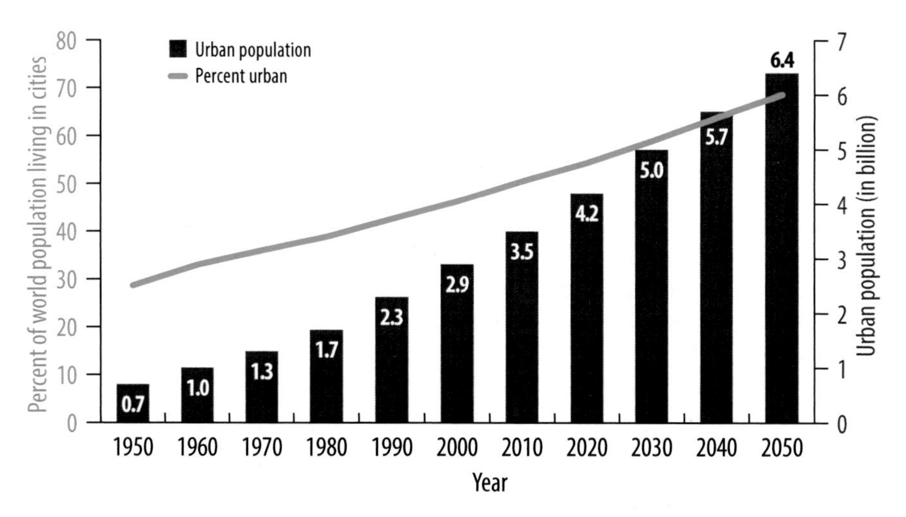

Fig. 1.10 The making of the urban age: in 100 years urban populations are projected to grow nearly tenfold, making up 70% of the global human population (Source: UN Department of Economic & Social Affairs, Population Division)

Modern cities of millions of people are certainly an astonishing achievement. They are the spaces in which humanity carries out the bulk of its social, economic and cultural transactions. They are the hubs of global communication and transport systems. They attract investors because they offer a vast variety of services at comparatively low per capita cost. One aspect of cities relating to enhanced sustainability is the already-existing empirical fact that urbanization is very positively correlated with reduced fertility rates.[76]

But there are also some ecological downsides: Urban resource demands and waste outputs represent a large part of human ecological footprints. Let us face up to a central contradiction: Cities are becoming our primary habitat, but urbanization in its present form is causing a rapidly increasing proportion of human ecological impacts. Studies from China and India have shown that people moving from a village to a city will, typically, increase their resource consumption fourfold.[77] The aggregated environmental impacts of humanity already vastly exceed the earth's carrying capacity (see Sect. 1.10).

Material affluence and urban sprawl go closely together. They are linked to people's desire for more living space, use of cars for commuting, and the wish to get away from urban noise, pollution and crime. Worldwide, urban growth and the transport infrastructures connecting cities are swallowing up ever more productive farmland. So the phenomenon of urbanization is increasingly also a problem for diminishing space for agriculture – and for wildlife. All this means that while cities are built on only a small proportion of the world's land surface, their ecological footprints now cover much of the productive land and sea surfaces of the globe.

---

[76] Martine et al. (2013).

[77] Sankhe et al. (2010). See also Brugmann (2009).

One co-author of this book, Herbie Girardet, established that London's ecological footprint is 125 times the surface area of the city itself, which is roughly the equivalent of England's entire productive land.[78] A typical North American city with a population of 650,000 would require 30,000 square kilometres of land, an area roughly the size of Vancouver Island, Canada, to meet its domestic needs. By comparison, a similar size city in India (with significantly lower living standards and a predominantly vegetarian diet) would require just 2800 square kilometres.[79]

The position of China, the world's most populous country, is particularly interesting: China has the world's most rapid urban growth, expected to rise from 54% in 2016 to 60% by 2020. Hundreds of millions of people have moved from village to city, and often, megacity. Recently, there has been much publicity about China's intention to create an ecological civilization (see Sect. 3.16). Of course, it is official government policy for urbanization to create agreeable prosperity. The 'National New-Type Urbanization Plan, 2014–2020',[80] indicates as much: 'Domestic demand is the fundamental impetus for China's development, and the greatest potential for expanding domestic demand lies in urbanization'. Both domestic demand and urbanization are also meant to reduce China's unhealthy (positive) trade balance. It has yet to be demonstrated that all this will not massively conflict with the country's environmental sustainability goals.

Is an urban world, dominated by sprawling cities and mega-cities, with their vast global ecological footprints, inevitable, or are there alternatives? Could cities exist and even thrive on regional rather than global resources? On a finite planet, could they be designed to continuously regenerate the resources they depend on? Section 3.6 will offer some optimistic answers.

1.8 Unsustainable Agriculture and Food Systems

Food security has been at the centre of all societies' concerns ever since humanity settled down and started to grow its food, rather than relying exclusively on hunting and gathering. Human ingenuity has seen global society moving from barely making it from one harvest to the next (often failing because of the weather, pests or other natural disasters), to a scandalous level of surpluses and waste.

While some 800 million people on earth still suffer from chronic hunger, about 2 billion are either overweight or obese, and another 300 million suffer from type 2 diabetes, all due to the inadequate quality and diversity of today's food supply and consumption patterns in both developed and developing countries. While the current agricultural system does produce surpluses, it also threatens our soils, water, biodiversity and, in fact, all ecosystems and their vital services, as well as the global climate.

[78] Girardet (1999).

[79] Int. Institute for Sustainable Development (IISD), as reported by www.gdrc.org.uem/e-footprints.html

[80] Chinese government. 2016. China to promote new type of urbanization. Feb 6, 2016. english.gov.cn

How did humanity get into such a situation, and what is needed to remedy it? Those questions have framed many recent studies of agriculture and the food system, including *Agriculture at a Crossroads,*[81] a groundbreaking report by the International Assessment of Agricultural Knowledge, Science and Technology for Development (IAASTD), commissioned by six UN agencies and the World Bank at the 2002 World Summit on Sustainable Development in Johannesburg. It was managed through a multi-stakeholder bureau with half government and half civil-society representatives. It occupied some 400 people, from farmers to scientists and experts in all agriculture and food-system-related disciplines, from all continents, for over four years.

This report was endorsed by 59 countries in March 2008. The key findings, although not fully supported by all the parties represented, were absolutely clear about the need for *a paradigm shift in agriculture and food systems*. These findings have been echoed by numerous further reports, including one by UNEP and the International Resource Panel, UNCTAD's 'Wake up before it is too late', and 'Smallholders, food security, and the environment' from the International Fund for Agricultural Development (IFAD).[82]

Farming plays a role in all the major dimensions of ecological damage. Destruction of biodiversity and disappearance of species are intimately connected with the ongoing clearing of forests and draining of wetlands, much of which occurs to obtain new farmland; agricultural fertilizer runoff disrupts the nitrogen and phosphorus cycles, causing dead zones in waterways; toxic pesticides and herbicides kill zillions of non-target animals and plants; and agriculture/forestry produces about 25% of the greenhouse gas emissions. So farming is one of the most crucial sectors that must change in order to alleviate the current ecological/climate crisis.

Industrial agriculture also displaces smallholder and indigenous farmers from their land. Smallholders make up a third of the world's population and half the world's poor; they nevertheless produce about 70% of its food on one quarter of its farmland,[83] and that mostly without the ecological damages listed above. Smallholder vulnerability is compounded by traditional customary forms of tenure which are frequently swept aside by national governments clinching corporate deals. Few of these dispossessions have prior consent or are properly compensated. Especially since 2006, 'land grabs' have accelerated, where corporations from the developed world, plus nations such as China and the Gulf countries, are taking over vast swaths of land, especially in Africa.

On a more generalized account, agriculture as it is done in our times turned out to be the most costly business with dramatic negative profit margins if the external costs

[81] Agriculture at the Crossroads. 2009. Washington: Island Press (One global Report, one executive Summary, and five regional reports.)

[82] UNEP and International Resource Panel. 2014. Assessing Global Land Use: Balancing Consumption with Sustainable Supply; UNCTAD. 2013. Trade and Environment Review 2013. Wake up before it is too late: Make agriculture truly sustainable now for food security in a changing climate; IFAD. 2013. Smallholders, food security, and the environment.

[83] GRAIN and La Via Campesina (2014). Big dams, industrial zones and mining also displace smallholders.

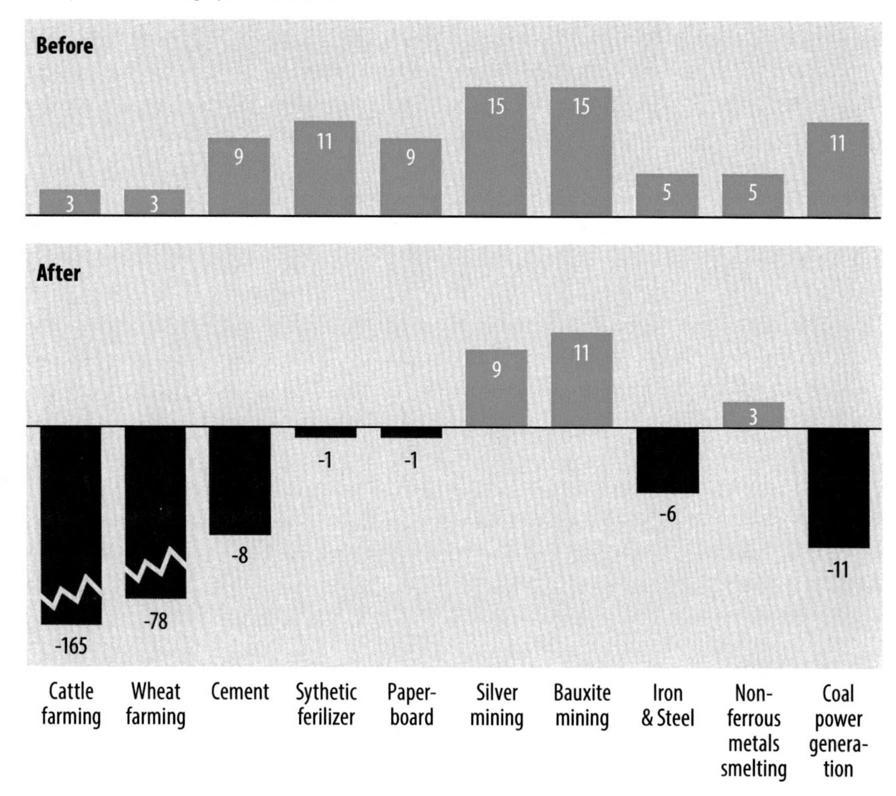

Fig. 1.11 Agriculture has by far the largest gap between superficial cost of production, transport and consumption on one side and 'true cost' on the other side (Data source: Trucost and TEEB 2013, courtesy Pavan Sukhdev)

are added to the mere production cost. Figure 1.11 shows ten different economic sectors. The first two, cattle farming and wheat farming, both core agricultural businesses, show by far the biggest 'losses' (brown bars) after inclusion of the 'natural capital cost', according to TEEB (The Economics of Eco-systems and Biodiversity).

The figure is startling. It shows that business activities in most sectors of the economy would be unprofitable – in fact show red figures – if the costs for using nature would be accounted for.

The IAASTD's in-depth analyses of the past 50 years concluded that, although there were some benefits in the short term, overall the Green Revolution of the 1960s failed to solve the key issue of hunger, which is lack of access to food rather than the overall supply. At the same time, large monocultures were preferred to increase farmers' labour productivity, and toxic chemicals became necessary to support high-yielding crops (many traditional staple crops tend to resist pests naturally). The new super varieties and hybrids were also very thirsty, and aquifers have been severely depleted. Because insect pests and weeds are very adept at developing resistance to poisons, after only a few seasons of use, many returned to pose increasing problems. Today, this same pattern is repeating itself with genetically modified crops.

Another unsustainable feature of modern agriculture, or rather of modern diets, is the steady increase of meat production and consumption. As Brian Machovina et al. are arguing, meat production is the single largest driver of habitat loss, and both livestock production and feedstock production are increasing in tropical developing countries, where the majority of biological diversity resides.[84]

Overall, it has been widely accepted, at least since the IAASTD Report of 2009, that 'business as usual' is not an option; if problems of climate, ecology, growing inequality and hunger are to be solved, agriculture needs radical transformation (see Sect. 3.5).

The oddities, inconsistencies, failures and destructive features of 'modern' agriculture are not well represented in the public media in any country of the world. The reason is simple. People want to eat and feel good at it, and farmers want to sell and feel good at it. The very idea that modern agriculture is deeply problematic is anathema to readers and listeners of the media.

What is a lot more popular in the media is the question if there is enough food around to satisfy all 7.6 and soon 10 billion people. The answer is not easy. A new compilation by R. Weiler et al. offers some important data and recommendations.[85]

Using up-to-date climatic, meteorological, geographical and demographic data, the authors arrive at the disturbing result that chiefly for Africa there will be a shortage of food towards the end of this century, owing to 'frightening desertification' and an expected massive increase of population. Apart from ethical considerations, the recommendations are mostly related to agricultural technology, but in many regards quite different from the views of the IAASTD.

1.9 Trade Versus Environment

One of the hottest international conflicts of our day is the debate about international trade. The Doha Round launched at the World Trade Organisation's (WTO) Ministerial Conference in Doha, Qatar, in 2001 has not led to any tangible results. It was meant to improve the trading prospects of developing countries, which felt that the radical moves towards global free trade adopted during the Uruguay Round of the GATT (Global Agreement on Tariffs and Trade), the WTO's predecessor, had chiefly benefitted the North and China. But neither the North nor the South seemed to be willing to

[84] Machovina et al. (2015).

[85] Weiler and Demuynck (2017).

come to an agreement on the Doha agenda. The North has been unwilling to abandon its agricultural export subsidies and the South is sceptical about benefits flowing to it.

The environment plays an absolutely marginal role in these trade negotiations. Most national regulations governing environmental protection are considered to be 'barriers to trade' and are therefore rejected. Along with bilateral and multilateral trade agreements, the WTO privileges cheaper production, market processes, monetary gain, business interests and economic growth. In 1991, for example, a GATT dispute panel ruled against the US ban on imports of tuna caught with collateral slaughter of dolphins because 'if the US arguments were accepted, then any country could ban imports of a product from another country *merely because the exporting country has different environmental, health and social policies from its own*'.[86] Here the WTO states baldly that trade has priority over environmental, health and social justice considerations, regardless of the wishes of a government and the people it represents. If the tuna catch destroys dolphins, that's too bad, but it is not relevant to trade.

Trade follows a different logic to the one that applies to the environment and consumer protection. The trade agenda, pushed first and foremost by the transnational corporations, is directed towards expansion of production and consumption, primacy of markets and the growth of private enterprise. It has no interest in issues of the public good (other than the possible benefits of supplying consumer items at a low price). It replaces 'rules for companies with rules for governments, and ... rules that protect consumers and the environment with rules that protect and facilitate traders and investors'.[87]

If a WTO dispute panel rules against a country, there are no palatable alternatives. It must either change its domestic laws, pay penalties representing 'lost profits' to the aggrieved corporation or face unilateral trade sanctions. The United States had to weaken its air pollution laws when the WTO ruled that it could not exclude petroleum from Mexico and Venezuela. Japan had to accept more pesticide residues in food than its own regulations demanded. In the dispute between Europe and the United States over growth-promoting hormones in beef, the WTO panel ruled against the EU and the United States was allowed to enforce retaliatory tariffs on various EU products.

For the WTO, objectors must prove harm rather than industry being required to prove safety. Europe, on the other hand, applies the 'precautionary principle' where novel substances are not permitted until the product is demonstrated to be safe on the basis of reliable scientific assessment of risk.[88] The leaks published by Greenpeace in May 2016 suggest that Europe's precautionary approach was going to be discarded in the planned Transatlantic Trade and Investment Partnership (TTIP).[89] Fortunately for European consumers and the environment, the resistance in the United States against the TTIP has grown with the advent of Donald Trump as President.

One nevertheless has to be cautious joining the choir of sovereignty advocates. In a highly interconnected world, negative environmental impacts tend to be global. The

[86] Emphasis ours. Higgs (2014), WTO (2010).

[87] Beder (2006).

[88] Higgs, op. cit. pp. 249–250. Sources are cited in full in this text.

[89] Neslen (2016).

Club of Rome therefore supports some kind of global governance that will limit the right of states to pursue such destruction. Global climate treaties are an example of such limiting rules. But so far, international trade treaties limit environmental rules. Trade rules are intended to enhance economic turnover, something that usually leads to environmental problems. It is ironic that the regulations made by the WTO are the only example of rules made at the global level where the legislation really has teeth. This is justified only if the WTO were obliged to give equal consideration to both the benefits and (environmental) hazards of free trade – which is absolutely not the case today (see Sect. 3.15 on global governance).

Meanwhile, in the absence of progress at the WTO, many countries have come to bilateral or multilateral trade agreements, filling the so-called spaghetti bowl of trade deals. The biggest such agreements – the trans-Pacific and trans-Atlantic agreements – were initiated by the United States during the Obama Administration. Although the Trans-Pacific Partnership (TPP) was signed in 2016, it won't be ratified by the US Congress. Similarly, the TTIP is very unlikely to be adopted.

US President Donald Trump has taken an openly protectionist stand, arguing that the loss of manufacturing jobs in the United States was the result of open borders allowing companies to seek cheaper human labour, lower taxes and weaker regulation (including on the environment). Here he joined the opposition to free trade that now exists in virtually all countries, with the exception perhaps of China and Singapore. This opposition reasons that free trade, while theoretically benefitting all partners, in reality serves as an invitation to companies to ignore the environment, human rights and the welfare of future generations. Those may not be President Trump's concerns, but motives do not always count when it comes to opportunistic alliances. What we are aiming at is an alliance for a fair balance between trade and public goods.

Almost by definition, free trade does help the strong and harms the weak; as the late Uruguayan journalist Eduardo Galeano put it, 'The international division of labour consists in this: some countries specialize in winning and some others in losing'.[90] While the official economic doctrine says that trade always serves both sides, the reality is less clear cut, and not just between countries. There are always losers in winning countries and winners in losing countries. The United Kingdom as a nation has pushed for more free trade for a long time, and the city of London has greatly benefitted from it. But the losers in Britain's traditional manufacturing regions got the upper hand in the Brexit vote, and they have blamed the EU (and the free movement of migrants), not their own government or global financial markets.

In the developing world, despite some success by China and the 'Asian tigers', many countries, especially in Africa and the Caribbean, saw local farmers and industries bankrupted by a flood of cheap imports. This was especially the case with agricultural products, since the United States and Europe have continued to subsidize exports from their farm sectors. Donald Trump's declared protectionism worries developing countries even more. As Martin Khor writes, Trump has shocked developing countries by considering tariffs on imports from developing countries

[90] Eduardo Galeano. 1973. (Spanish original 1971) Open Veins of Latin America: Five Centuries of the Pillage of a Continent SKU: mrp9916, Paperback ISBN: 9780853459910.

where the United States has a trade deficit,[91] and also by reducing UN funding, to the detriment of social and environmental programmes in many developing countries. Khor also mentions Trump's blatant disrespect for the environment, and his likely withdrawal from international environmental treaties and conventions.

Another aspect of trade is the intensified global flow of capital which the WTO has fostered by limiting the rights of governments to regulate the entry, behaviour and operations of foreign-based corporations. Although this has major effects on the environment, that is not the reason why people tend to be concerned. After the global financial crisis of 2008, a UN panel chaired by Joseph Stiglitz pointed to numerous problems with financial liberalization. This UN panel of experts recommended that 'agreements that restrict a country's ability to revise its regulatory regime – including not only domestic prudential but, crucially, capital account regulations – obviously have to be altered, in light of what has been learned about deficiencies in this crisis'.[92] Alas, the panel's recommendations have not been adopted by the WTO.

One crucial bias embedded in financial deregulation was captured by Indian economist Prabhat Patnaik, who wrote that the local financial sector has been detached from its 'anchorage in the domestic economy to make it part of the international financial sector; … and to remove it from the ambit of accountability to the people'.[93] This aspect of 'free trade' gives the financial markets a dangerously dominant power over investment worldwide. No considerations of local interests, public good or democratic control can apply.

To sum up, trade is a good thing and occurs only if both sides expect benefits from it. But trade is also a segment of international competition, which can lead to the defeat of weaker companies or entire states, and trade has enormous effects on natural resources and the environment in general but so far lacks appropriate rules protecting those public goods. The world will need a new appreciation of balance (see Sect. 2.10). Regarding trade, this will mean a level playing field between commercial and environmental objectives.

1.10 The 2030 Agenda: The Devil Is in Implementation

Three months before the Paris climate agreement, the United Nations oversaw another unanimous agreement: the *2030 Agenda*,[94] consisting chiefly of 17 *Sustainable Development Goals* (SDGs) as well as 169 targets to specify the SDGs. Figure 1.12 offers pictograms of the 17 goals.

---

[91] Khor (2017).

[92] UN (2009).

[93] Patnaik (1999).

[94] Full title: Transforming our world: the 2030 Agenda for Sustainable Development A/69/L.85 – Draft outcome document of the United Nations summit for the adoption of the post-2015 development agenda.

Fig. 1.12 The 17 Sustainable Development Goals of the 2030 Agenda. SDGs 1–11 can be considered as the socio-economic goals. SDG 12 is about responsible (sustainable) consumption and production; SDGs 13–15 are environmental goals. SDG 16 is on peace, justice and institutions, SDG 17 on partnerships in the process

The declaration accompanying the SDGs contains a vision statement that includes the statement '… we envisage a world in which development and the application of technology are climate-sensitive, respect biodiversity and are resilient. One in which humanity lives in harmony with nature and in which wildlife and other living species are protected'.[95]

While the Club of Rome lends strong support to this 'supremely ambitious and transformational vision', there remains a need to examine the consistency of the SDGs and the modalities under which the goals will be implemented. What is really the meaning of the quoted statement. It surely relates to the three environmental SDGs, speaking in affirmative language about urgent action needed to combat climate change (Goal 13); the importance of conserving and sustainably using the oceans, seas and marine resources for sustainable development (Goal 14); and protecting, restoring and promoting sustainable use of terrestrial ecosystems, sustainably managing forests, combatting desertification, halting and reversing land degradation, and halting biodiversity loss (Goal 15).

Nowhere, however, is it admitted in the 2030 Agenda that the successes in reaching the eleven social and economic goals (Goals 1–11), *if done based on conventional growth policies*, would make it virtually impossible even to reduce the speed of global warming, to stop overfishing in the oceans or to stop land degradation, let alone to halt the loss of biodiversity. In other words, assuming no major changes in the way economic growth is defined and pursued, humanity would be confronted with massive trade-offs between the socio-economic and the environmental SDGs.

[95] Transforming our world, l.c. p. 3.

If the fate of similar objectives mentioned in the Earth Summit's Agenda 21[96] is any guide, solutions to the socio-economic deficits will be tackled by attempting to accelerate growth and trade, leading to a cascading erosion of the environment, whether climate, oceans or terrestrial systems. Again going on the past 25 years, the degree of socio-economic progress this tactic is likely to achieve will still be far less than what is really required. A radical new synthesis will be needed.

This synthesis has to acknowledge that for developing countries the conflicts between social and environmental objectives are often muted. The developing world frequently refers back to the powerful slogan coined by the late Indian Prime Minister Indira Gandhi when she attended the first UN environment summit, in Stockholm 1972. Her slogan was 'poverty is the biggest polluter'. At that time, the statement had much truth to it. Environmental issues were mostly local pollution problems, and the evident answer was pollution control – costing money which only the rich can afford.

The trouble is that, in our day and age, a more accurate slogan should be 'affluence is the biggest polluter'. This is because greenhouse gas emissions, resource consumption as well as land-use change destroying soil quality and biodiversity-rich habitats are companions of affluence. This reality is clearly seen in the recent report by Chancel and Piketty,[97] who trace global inequalities in carbon emissions in the period 1998–2013. They note that the three million wealthiest Americans (the top 1%) have on average CO_2 emissions in the range of a staggering 318 tons per capita per year, while the world average per person is only around 6 tons! So more than 50 times the pollution and use per rich person compared to the average person, let alone to the poorest people on earth.

It is often said that it is useless to be concerned about the conspicuous lifestyles of the rich, the simple reason being that they are so few. But Piketty's data tell a different picture. The fact is that the 1% richest Americans account for roughly 2.5% (!) of *global* greenhouse gases. If the top 10% richest households in the world are targeted, their contribution to GHG emissions would make up 45% of the total. So the real bang for the buck is to change the habits of the rich, not the poor.

This means that developing countries are right in saying that the biggest burden of changing course should be on the affluent nations. Clearly, developing countries see it as their own priority to pursue the socio-economic SDGs, like poverty eradication (Goal 1), food security (Goal 2), health (Goal 3), education (Goal 4) and employment for all (Goal 8). After all, these goals are meant to apply to all human beings in the world – 7.6 billion today, 9 billion in less than 20 years and perhaps 11.2 billion towards the end of this century.[98] This is a nightmare figure assuming that the world is not willing or not able to change course in fertility habits (see Sect. 1.7).

---

[96] 80 United Nations Conference on Environment and Development (1992).

[97] Chancel and Piketty (2015).

[98] The world's population is expected to reach 8.5 billion by 2030 and 9.7 billion in 2050, a new United Nations report says. And there should be 11.2 billion people on Earth by the end of this century. Source: associated Press, 29 July, 2015.

As long as 'affluence remains the biggest polluter', the mentioned trade-offs between the socio-economic and the environmental goals in the SDGs will prevail and will ultimately overshadow and destroy the success of the socio-economic goals. On the other hand, everyone will agree with the UN statement that 'the 17 SDGs are universal goals and targets which must include the entire world, developed and developing countries alike. These goals are integrated and indivisible and balance the three dimensions of sustainable development'.[99]

Recent studies seem to confirm that the trade-offs between socio-economic and ecological SDGs are indeed major. Arjen Hoekstra's study on water footprints[100] indicates that achieving food security (Goal 2) can easily conflict with providing enough water for all (Goal 6); the effects on biodiversity (Goal 15) are as yet unaccounted for but are massive and nearly all negative. The International Resource Panel did a preliminary assessment of interlinkages and trade-offs between different SDGs,[101] finding that a large number of goals for human well-being (11 of the 17) are 'contingent on the prudent use of natural resources'. This is a very diplomatic way of saying that achieving the socio-economic goals while applying the prevailing non-prudent use of natural resources is simply impossible. In parallel, Michael Obersteiner et al.[102] found massive trade-offs between policies to lower food prices and policies advancing SDGs 13, 14 and 15.

Of course, it would be unfair and one-sided to criticize the socio-economic goals (with formulations stemming mostly from the developing countries) without addressing and criticizing the overconsumption of the rich of this world. Even when ecological destruction takes place in the developing world, it often occurs in the context of harvesting or manufacturing exports that end up servicing the affluent. The developed world outsources much of the environmental damage involved in its consumption patterns – about 30% of all species threats, for example, are due to international trade.[103] The Club of Rome has always stood for the principles of justice and fair distribution. This means that, when addressing the trade-offs between economic and ecological SDGs, we should always look for solutions that embody North-South justice.

In a recent study, Jeffrey Sachs et al.[104] offer some quantitative assessment of the performance and challenges in achieving the SDGs at the present time. Using existing indicators provided by the World Bank and other institutions, countries were assessed on indicators for each goal and were ranked according to their overall performance across all 17 SDGs. Figure 1.13 shows the top ten performers and a few other major countries.

[99] From para 5 in the document referred to in footnote 66.

[100] Hoekstra (2013).

[101] International Resource Panel and Development Alternatives (Lead author: Ashok Khosla) 2015. Addressing Resource Inter-linkages and Trade-offs in the Sustainable Development Goals. Nairobi.

[102] Obersteiner et al. (2016).

[103] Lenzen et al. (2012).

[104] Sachs et al. (2016).

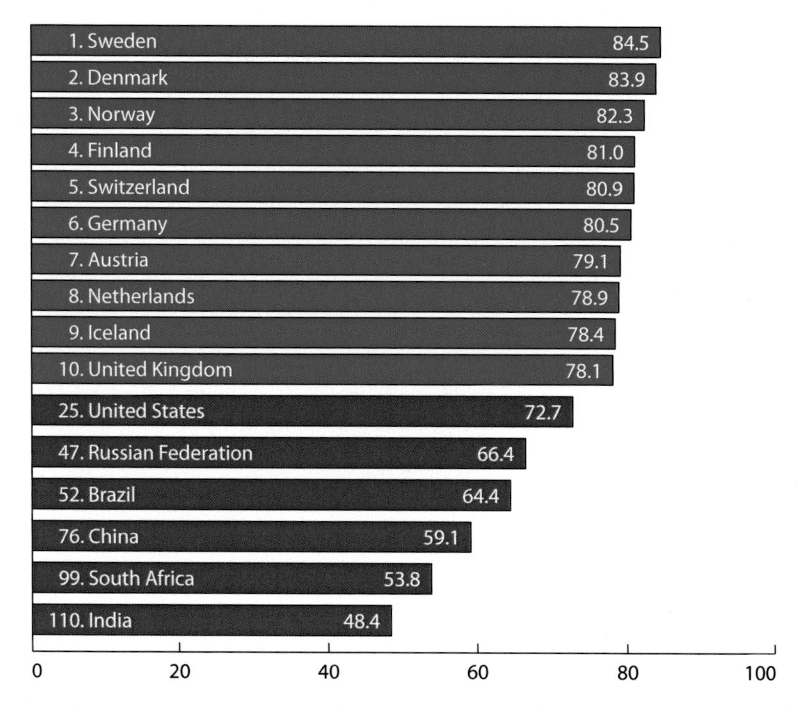

Fig. 1.13 Country ranking by current SDG performance (composite index; maximum score is defined as 100.) The first ten countries are all European countries (and Iceland). The United States is lagging behind due to high inequality and excessive resource consumption. Developing countries remain weak because of high levels of poverty, hunger, illiteracy and high levels of unemployment (Source: https://www.bertelsmann-stiftung.de/en/topics/aktuelle-meldungen/2016/juli/countries-need-to-act-urgently-to-achieve-the-un-sustainable-development-goals)

Strikingly, the first ten countries are all prosperous European countries while the ten lowest ranking countries (see table below) are all poor and mostly African. The lowest 10 among the 149 countries measured are as follows:

Rank	Country	Performance
139	Afghanistan	36.5
140	Madagascar	36.2
141	Nigeria	36.1
142	Guinea	35.9
143	Burkina Faso	35.6
144	Haiti	34.4
145	Chad	31.8
146	Niger	31.4
147	Congo, Dem. Republic	31.3
148	Liberia	30.5
149	Central African Republic	26

At first glance, such figures are not too surprising. The 2030 Agenda is meant to lift poor countries to much higher levels. At second glance, however, there is one

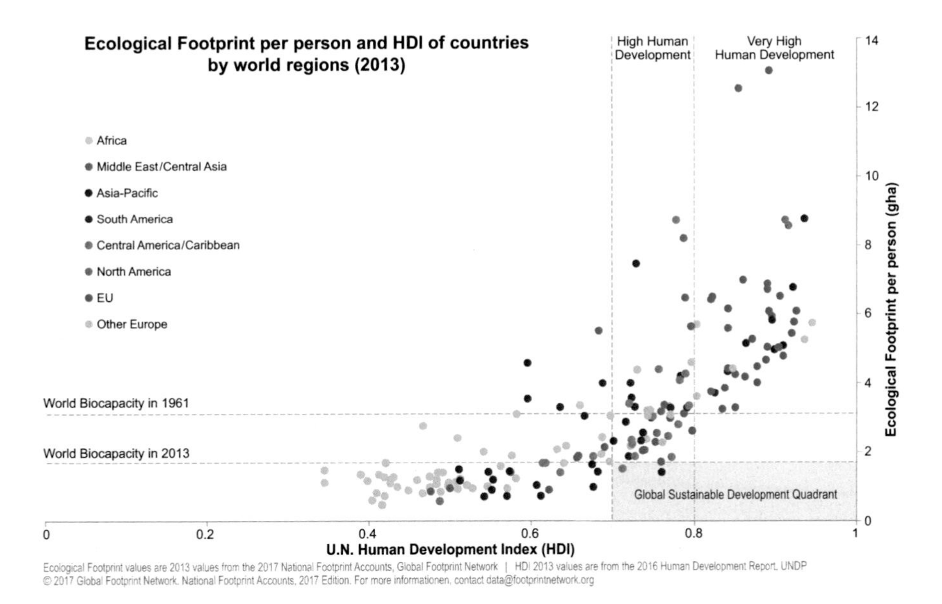

Fig. 1.14 Sustainability chart according to the Global Footprints Network. Per capita ecological footprints from bottom to top (hectares per person), and Human Development Index (HDI) from left to right. Poor countries (*left*) have deplorably small HDI, and rich countries have deplorably large footprints – leaving the 'Global Sustainable Development Quadrant' nearly empty. The upper dotted line shows the per capita world biocapacity in 1961 at a world population of 3.1 billion (Source: 2017 Global Footprint Network. National Footprint Accounts, 2017 Edition; data.footprintnetwork.org)

disquieting fact about this study: It reveals that high SDG performance strongly correlates with the conventional development path of growth, including the overuse of natural resources as measured by per capita ecological footprints.

A country's ecological footprint, annually assessed and updated by the Global Footprint Network, measures the area required to supply the goods and services consumed by its population. Not surprisingly, this measure is generally higher for countries with high socio-economic performance and affluence.

Figure 1.14 shows the per capita ecological footprint in the SDG-ranked countries (vertical axis) plotted as a function of the average Human Development Index, HDI (horizontal axis), of the people in the respective countries.

HDI is a composite indicator of education, health and income per capita, which is used to gauge the well-being of people in different countries. In the lower right corner of the figure, the 'Global Sustainable Development Quadrant' is where the HDI is above 0.8 and the per capita footprint is below 1.8 hectares.

The disquieting fact is that the sustainable development rectangle is almost empty, meaning that there is no single country that shows a high socio-economic performance (HDI above 0.8) and at the same time achieves sustainable scores (below 1.8 hectares) on the footprint measure. Translated into the SDG agenda this means that there is no single country with a high performance on all three 'pillars' (economic, social, environmental).

Sachs et al. therefore reveal a hidden paradox: If all 11 or 12 socio-economic SDGs were achieved in all countries, one would expect average footprints to reach

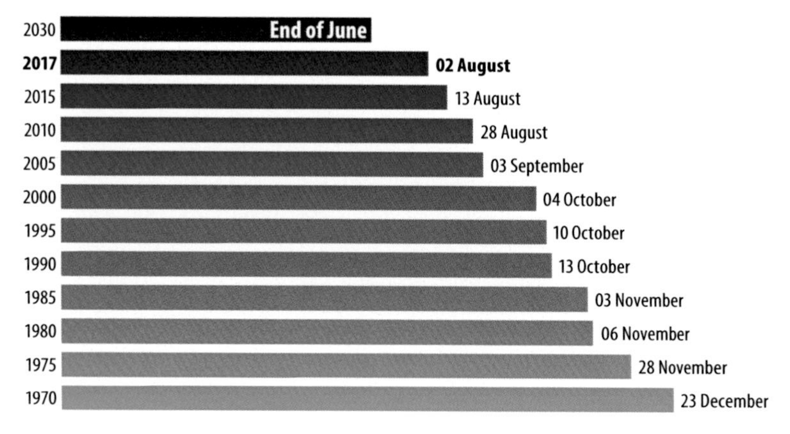

Fig. 1.15 The 'overshoot day' is moving up the calendar (Source: www.overshootday.org)

sizes of 4–10 hectares per person. For 7.6 billion people, that would mean we would need between two and five planets of the size of the earth!

Put into another impressive figure, the ecological footprint allows for the estimation of the 'overshoot day', the day after which the world begins consuming resources that will not be replenished during the rest of the year. Whereas in 1970 it was in late December, in 2017, this day had already moved to August 2 and it is expected to be as early as June by 2030 (Fig. 1.15).

Sachs et al. stress that even the frontrunner SDG countries are far from being ecologically sustainable.

In summary, one would conclude from the discussion of the SDGs adopted by the United Nations that the world cannot possibly afford to pursue those 17 goals separately. A *coherent* policy will be needed to address socio-economic and environmental goals as a whole. This, however, will force the world to fundamentally overhaul the technological, economic and political approach to development as it has been practised for many decades.[105]

1.11 Do We Like Disruptions? The Case of the Digital Revolution

1.11.1 Disruptive Technologies: The New Hype

Technological innovations and development are speeding up. In America, innovation is what (nearly) everybody is aiming at. The new term garnering real excitement, however, is 'disruptive technologies'. It means innovation that replaces and destroys the existing technology, for example, smart phone cameras replacing traditional photography (Kodak, once a highly profitable company, went bankrupt in a

[105] This is also the view of Michael Wadleigh and Birgit van Munster. 2017. Nature perspective, closed mass Homo Sapiens Foundation. hsfound@gmail.com, closedmass@gmail.com

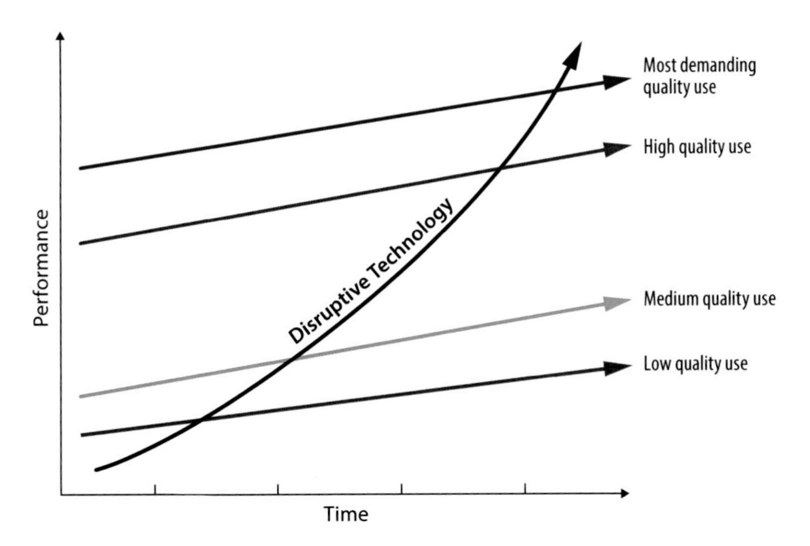

Fig. 1.16 Disruptive technology may start below low-quality use or standards but eventually overtakes even most demanding standards due to its dynamic of creating or conquering new markets. Schematic picture from the Wikipedia entry about disruptive technology (Accessed July 24, 2016)

few years; or music streaming replacing CD records). The term was coined by Clayton Christensen and published in 1995 by Bower and Christensen.[106] The concept is visualized in Fig. 1.16.

Until 1995, the connotations of *disruptive* were negative. Do you like to be 'disrupted' when sleeping, making love or enjoying dinner with friends? Most readers probably don't. But for innovation addicts this is the real excitement. The authors of *Disruptive Technology* refer to Joseph Schumpeter's notion of *creative destruction.* Schumpeter in 1942[107] shocked his readers by giving *destruction* a positive meaning: The 'good' innovation surpasses and thereby destroys old structures and technologies. He called it 'the essential fact about capitalism'. Despite origins in Schumpeter's thought, Bower and Christensen naturally did not want to call their brainchild *destructive* technology. Conveniently, the adjective *disruptive* was still available with not too much of a negative meaning. But in this chapter we cannot avoid – for all our admiration for ingenious and successful technological innovations – also looking at the dark sides of destruction and disruption.

1.11.2 Digitization Is the Buzzword of Our Time

Today, a tremendous acceleration of technological innovation can be observed. Digitization is the buzzword of the time. The young see themselves as 'digital natives', and look down a bit on the 'digital immigrants', the elderly who grew up with books

[106] Bower and Christensen (1995).

[107] Schumpeter (1942).

and pens and paper. The behaviour of the digital natives keeps changing rapidly, in line with thousands of new apps every year and indeed with the digitization of our society. And they usually enjoy the disruption they experience.

People devote a significant part of time, attention and resources to digital artefacts. While there are many other domains where technology is evolving, digital has become something of a synonym for 'technology' and a dominant part of the public sphere. Technological innovation is speeding up and introducing new products and services, altering processes, shaking markets and ultimately changing our lives, by inducing transformations that are deemed 'disruptive' – using the positive meaning of the term.

Since the 1980s, an explosive growth has occurred in information and communication technologies (ICTs), and their presence has become pervasive. The widespread frenzy provoked by the latest digital gadget mirrors an exciting entrepreneurial spirit which is mobilized by the potential of technologies to address human desires. In parallel with the explosion of ICTs, however, humanity has become more aware of the many and intertwined challenges it faces to make life in this planet enjoyable and sustainable in the long run.

The Brundtland Commission popularized in 1987 the concept of 'sustainable development' (SD) almost in sync with the launching of the first personal computers (IBM PC in 1981, Commodore 64 in 1982 and Macintosh in 1984). But in the meantime considerable negative impacts, both social and ecological, of the digital revolution have become apparent.

The size and speed of the digital transformation is unprecedented. It will require all kinds of human capacities to respond and live with it. The best and brightest of researchers and innovators should engage in responding to the challenges. Some could explore how best to use digital technologies to overcome the downsides of our unsustainable way of life.

What's next? It is unclear if 'blue oceans', that is, uncontested market space between companies allowing new services, will continue to be discovered, as has been the case for decades in the world of IT. In the meantime, new assailants try to create their own 'blue oceans' more often by using digital tech to bypass current regulations, labour arrangements and fiscal systems. Under the slogan of 'zero marginal costs', they principally seek to evade taxes. Tax-paying taxi drivers are being dispossesed by Uber, which avoids the full costs of transport and minimizing tax payments while creating a new monopolistic brand. The concept of a 'sharing economy' is certainly an appealing one, but it needs the appropriate framework to ensure that its business companies also share the cost of infrastructure – by paying appropriate taxes at the place where they earn money.

One of the most talked-about trends of the day is 3D printing, which is marketed as a means of empowering citizens. It is supposed to bring to any of us the capacity for self-production at home, with easy access to new eco-friendly designs inspired by nature and requiring less energy and raw materials, with improved durability, weight and efficiency. 3D printing is impressive, but still has to pass the reality check from economic, societal and *ecological* points of view.[108] Imagine only the

[108] Vickery (2012).

supply of feedstock. If millions of decentralized 3D printers need a steady supply of anything between 20 and 60 different chemical elements (and more compounds), one would expect an explosive increase in demand for and massive distribution of those chemicals. And the recycling of chemical elements used in milligrams remains a nightmare.

1.11.3 Scary 'Singularity' and 'Exponential Technologies'

Jeremy Rifkin is one of the early proponents of a new economy – in his words a Third Industrial Revolution[109] – that will emerge as the consequence of a set of new and disruptive technologies, underpinned by ICT. His vision may be a bit narrow, essentially focussing on renewable energy and its decentralization powers. In reality, the new Industrial Revolution goes far beyond that.

As a matter of fact, Rifkin's 'third' Industrial Revolution is closely related to what is nowadays referred to as the Fourth Industrial Revolution, usually called Industry 4.0. In this chapter, emphasis is laid on the more frightening side of that revolution. Emphasis on the positive sides follows in Chap. 3.

From a technical point of view, two main drivers are at the core of the process of digitization. The first is Moore's law (named after the founder of Intel) which holds now for more than 40 years and states that technical progress in miniaturization makes it possible for the number of transistors in a dense integrated circuit to approximately double every 2 years. This has enabled the computing power of microprocessors to be increased extremely fast without increasing their cost.

The second driver is Metcalfe's law, stating that the value of a network is proportional to the square of the number of connected users. This means that a competitive diffusion process over a network can be very fast because the advantage of the leading player is more than linear; it is quadratic. Software businesses, telecommunications and the Internet exhibit such strong positive network feedbacks.

These observed characteristics are now used as foundations for a new belief in 'exponential technologies'. The implication seems to be 'exponential innovation' as a process able to disrupt all areas of human practices for our benefit. Ray Kurzweil and Peter Diamandis are the best known promoters of this vision of infinite improvements, which they interpret as *the way to a new world of abundance*,[110] in which all the needs of the soon-to-be ten billion inhabitants of the planet will be met by the use of new and fascinating technologies of water purification, food production, solar energy, medicine, education and the reuse or recycling of rare minerals. In stark contrast with the mostly 'linear-thinking executives' of major corporations all over the world,[111] a small group of 'exponential entrepreneurs' are expected to find solutions to the big problems by exploiting the cycles of '6 Ds': digitization, deception (until enough growth is achieved), disruption, demonetization, dematerialization and democratization.

[109] Rifkin (2011).

[110] Diamandis and Kotler (2012).

[111] Diamandis and Kotler (2015).

Here comes one of the scary points. Peter Diamandis and Steven Kotler don't seem to be familiar with the 'rebound effect', which essentially says that in the past, all efficiency increases created higher availability of the desired products, leading invariably to higher consumption and in consequence to rising ecological damages, such as global warming, resource depletion and biodiversity losses (often caused by intensified human transport).

And there are societal consequences. One has been turned into a novel. Dave Eggers in *The Circle* shows how the powers of the world's biggest Internet company can become overwhelming. The situations resemble those of George Orwell's *1984*, if in a funnier language and closer to today's reality.[112] However outlandish these fears may seem at the moment, one should not be naïve. The digital world – as well as other parts of the business community – facilitates the emergence of monopolies including gangster conglomerates.

What is scarier still is Ray Kurzweil's vision of 'Singularity',[113] when 'artificial intelligence' will surpass human, from which point on an accelerated speed of 'innovation' occurs. Readers are invited to reflect for a moment how the dynamics of self-accelerating innovations created by supercomputers can be controlled. The genie will have left the bottle. And then combine that uncontrollability with the prospects of modern high tech weapons, hysterical or misinformed leaders, and people's ignorance of the laws of physics.

Another consideration is the excitement with *exponential technologies*, cultivated at the 'Singularity University' in Sunnyvale, California. Peter Diamandis serves as the president of this high tech think tank, which propounds the idea of continuous, exponential growth in technology and innovation.

Good science proves that resource-related exponential phenomena are viable only for limited periods of time. In the case of closed systems such as bacteria on a Petri dish, after the slow 'lag phase' comes the exponential 'log phase', followed by the stationary phase. And that tends to lead into the 'death phase', as the bacteria exhaust their own resource base.

There are certainly differences between biology and electronics, but in stark contrast with the arrogant optimism of the Singularity vision, the industry-sponsored International Technology Roadmap for Semiconductors (ITRS) now recognizes that Moore's law will not hold forever, that its dynamics will fundamentally change around 2020 or 2025 because of physical limits and of the challenge of controlling heat emissions at microscopic level.[114] So miniaturization of transistors seems close to its end. Maybe our civilization should be humbler about the prospects of exponential innovation, after all.

For all the good things attributed to ICTs and digital technologies, when considering their direct impacts in terms of sustainability, there is no doubt that the first-order effect is negative. The ICT sector itself has led to a rapid, in many cases exponential, increase in the use of energy, water and some critical resources, like

[112] Eggers (2011).

[113] Kurzweil (2006).

[114] Suhas Kumar (2015).

specialty metals. This is not the place to get into much details, but the evidence is accumulating and has many different faces. Readers may look up some of the references.[115,116,117,118,119]

1.11.4 Jobs

One of the biggest concerns relating to disruptive digital innovation is associated with the elimination of jobs. Politically, this is extremely sensitive. Actually, new digital outfits dream of replacing employees by robots. So the danger is apparent of a general disappearance of jobs, a question widely discussed for several years. A frequently cited study by Carl Benedikt Frey and Michael Osborne shows that 47% of jobs (in the United States) are at risk of automation, as highlighted in Fig. 1.17.[120] A 2016 report by the World Economic Forum[121] concludes that about 7.1 million jobs will be lost and 2 million jobs will be created in 15 important countries over the next 5 years, with a net loss effect of 5.1 million jobs. Newly industrialized countries with a still underdeveloped technological infrastructure are likely to be more negatively affected than some of the old and rich industrialized countries. Labour-intensive industries producing parts for major manufacturers located in rich countries are vulnerable as well.

More dramatic figures can be found in many places. To quote only one: A recent ad says that 'By 2020, the global economy is set to have a shortfall of 85 million skilled jobs'. The ad sponsored by Chevron and the 49ERS Foundation continues with an educational remedy strategy saying, 'In the next decade, 80% of all professions are expected to require STEM skills' (STEM standing for science, technology, engineering and mathematics).[122]

Of course, the decline and disappearance of traditional jobs – due to automated production and other types of digitization – should be an impulse for the creation of new jobs related to education and care, and especially to activities required by a massive transition to sustainability. But such jobs traditionally depend mainly on public sector initiatives and public sector finance. How will that happen in an economic system where raising taxes seems to be a non-starter?

Adding to the job fears is the fact that digital disruption also means worse labour relations, de-unionized and based on low-cost labour except for a fairly small elite of techies.

---

[115] European Union (2014).

[116] Williams et al. (2002).

[117] Silicon Valley Toxics Coalition (2006).

[118] Hintemann and Clausen (2016).

[119] Climate Group for the Global eSustainability Initiative (2008).

[120] Frey and Osborne (2013).

[121] World Economic Forum (2016).

[122] TIME (2017).

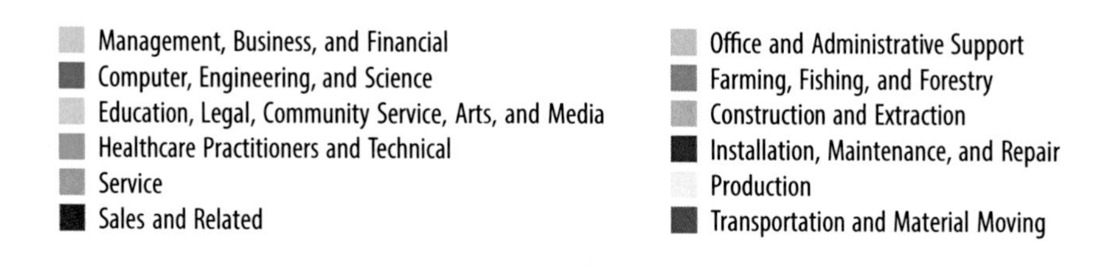

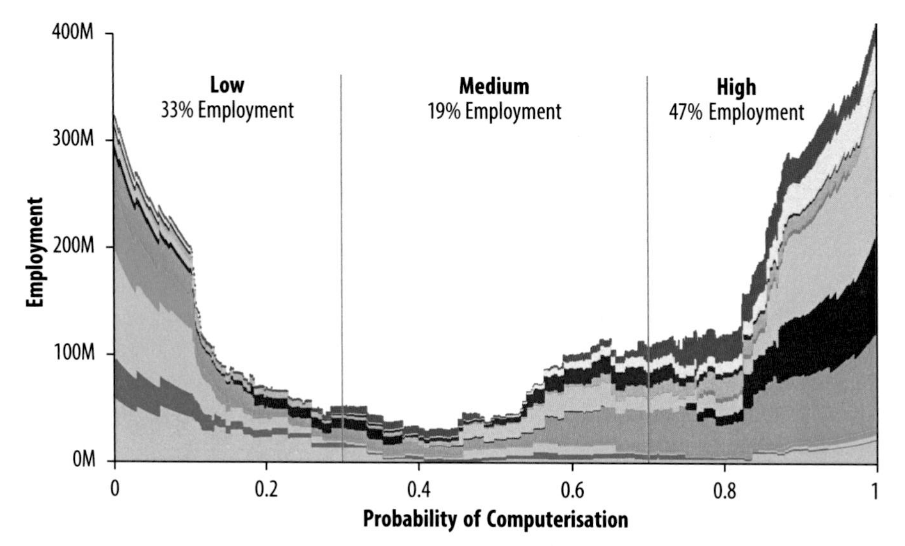

Fig. 1.17 Probability of jobs being lost through computerization or digitization. Forty-seven per cent of (US) jobs have a more than 70% probability of being lost (Source: Frey CB, Osborne MA (2016) The future of employment: How susceptible are jobs to computerization? http://www.sciencedirect.com/science/article/pii/S0040162516302244)

1.12 From Empty World to Full World

Among economists and high officials in government, one often hears the statement 'There is no conflict between economics and ecology. We can and must grow the economy and protect the environment at the same time'. Is this true? Is it possible? Although it is a comforting idea, it is at most half true.

Given the issues dealt with so far, it is natural for the Club of Rome to conclude Chap. 1 of the book with a discussion on economics, primarily by highlighting the huge difference between an empty world and a full world. The principles guiding our economies in a full world ought to be very different than in an empty world.

1.12.1 The Impact of Physical Growth

The human economy, as shown in Fig. 1.18, is an open subsystem of the larger ecosphere that is finite, non-growing and materially closed, although open to a continual throughput of solar energy. When the economy grows in physical dimensions, it incorporates matter and energy from the rest of the ecosystem into itself.

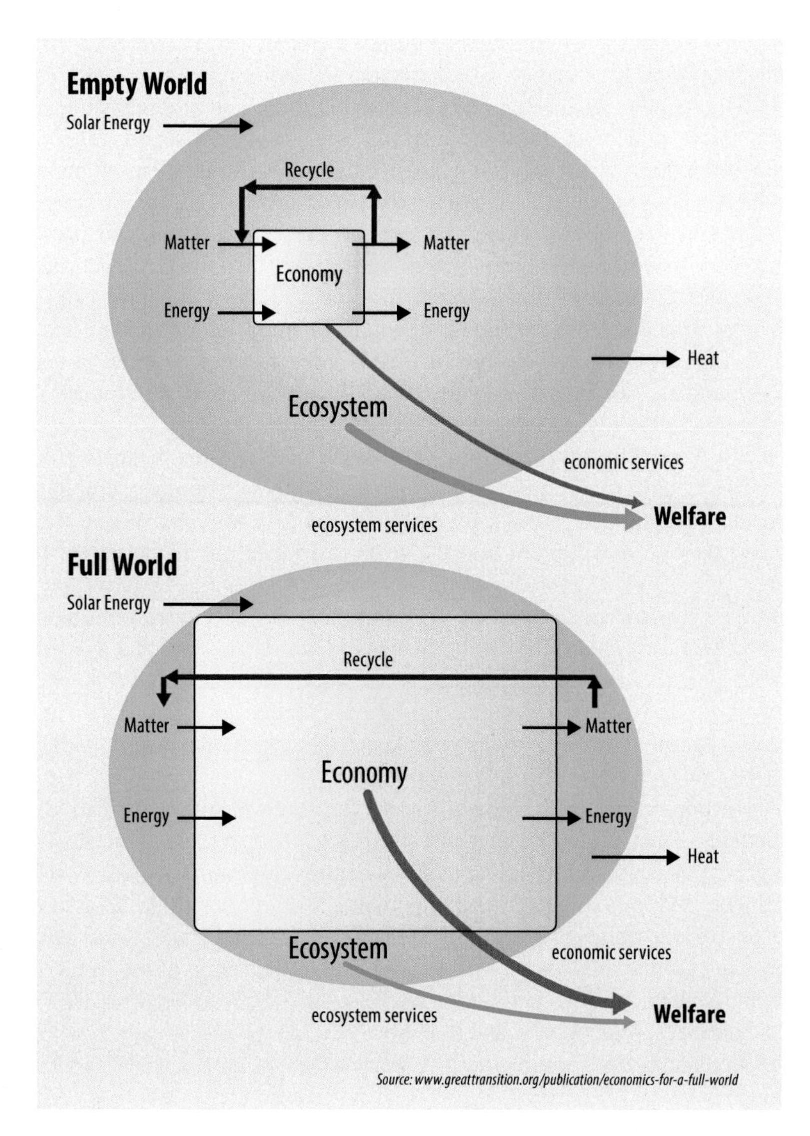

Fig. 1.18 Welfare in a full versus empty world (Source: Herman Daly, www.greattransition.org/publication/economics-for-a-full-world)

That means that what is called 'the economy' must, by the law of conservation of matter and energy (the first law of thermodynamics), encroach on the ecosystem, diverting matter and energy from previous natural uses. More human economy (more people, commodities and waste dumps) means less natural ecosphere. There is an obvious physical conflict between the growth of the economy and the preservation of the environment.

That the economy is a subsystem of the ecosphere seems perhaps too obvious to emphasize. Yet the opposite view is common in our governments. For example, the chairman of the UK Natural Capital Committee says that 'As the White Paper rightly emphasised, the environment is part of the economy and needs to be properly integrated into it so that growth opportunities will not be missed'.[123]

But how crucially important is this conflict between how physicists understand the laws under which the planet exists and what economists and governments believe? Some think not at all. Some believe that we still live in an empty world, where the economy was small relative to the containing ecosphere (relatively empty of humans and our things), where our technologies of extraction and harvesting were not very powerful, and our numbers were small. Fish reproduced faster than we could catch them, trees grew faster than we could harvest them, minerals in the earth's crust were concentrated and abundant, and natural resources were not really scarce. In the empty world, the unwanted side effects of our production systems, which economists call 'negative externalities', were dispersed across vast natural landscapes and were often absorbed with little impact.

In the full world, however, there is no vast natural sink to absorb wastes. Carbon dioxide accumulation in our atmosphere today is a salient example. In the full world, 'externalities' are not external but affect people and planet alike. By definition, they are not figured into the costs of production as the expenses that they are.

Both Neoclassical and Keynesian economic theories developed on the basis of the empty-world vision and still embody many assumptions from that past era. But remember Fig. 1.6: In one lifetime the world population has more than tripled – from two billion to over seven billion. And the populations of cattle, chickens, pigs, soybean and corn stalks have grown even faster, as have the non-living populations of cars, buildings, refrigerators and cell phones.

All these populations, both living and non-living, are what physicists call 'dissipative structures'. That is, their maintenance and reproduction requires a metabolic flow, a throughput that begins with depletion of low-entropy (high structure) resources from the ecosphere and ends with the return of polluting high-entropy (high disorder) waste right back to the ecosphere. At both ends, this metabolic throughput imposes costs that are necessary for the production, maintenance and reproduction of the stock of both people and wealth. Until recently, the concept of metabolic throughput was absent from standard economic theory, and even now its importance is greatly downplayed, in spite of the important contributions of Nicholas Georgescu-Roegen[124] and Kenneth Boulding.[125]

The costs and benefits of the transition from empty to full world are shown in Fig. 1.18. The brown arrow from Economy to Welfare represents economic services (benefits from the economy). It is small in the empty world but large in the full world. It grows at a diminishing rate (because rational beings satisfy their most important wants first – law of diminishing marginal utility). The costs of growth are represented by the shrinking ecosystem services (green arrow) that are large in the

[123] Dieter Helm, Chairman of the Natural Capital Committee, The State of Natural Capital: Restoring our Natural Assets, UK. 2014.

[124] Georgescu-Roegen (1971).

[125] Boulding (1966).

empty world and small in the full world. It diminishes at an increasing rate as the ecosystem is displaced by the economy (because humans – *at best* – sacrifice the least important ecosystem services first – law of increasing marginal costs).

Total welfare (the sum of economic and ecological services) is maximized when the marginal benefit of added economic services is equal to the marginal cost of sacrificed ecosystem services. As a first approximation this gives the optimal scale of the economy relative to the ecosphere. Beyond this point physical growth costs more than it is worth and thus becomes *uneconomic growth*. The empirical difficulty of accurately measuring benefits and costs (especially costs) should not be allowed to obscure the logical clarity of an *economic limit to growth* – or the impressive empirical evidence of the same from the Global Footprint Network and the planetary boundaries study.

Recognizing the concept of metabolic throughput in economics brings into play the laws of thermodynamics, which is inconvenient for the 'growthist' ideology. The first law of thermodynamics, as noted above, imposes a quantitative trade-off of matter/energy between the environment and the economy. The second law of thermodynamics imposes a qualitative degradation of the environment – by extracting low-entropy resources and returning high-entropy wastes. The second law thus imposes an additional conflict between expansion of the economy and preservation of the environment, namely, that the order and structure of the economy is paid for by imposing disorder and destruction in the sustaining ecosphere.

1.12.2 *The GDP Fallacy: Physical Impacts Ignored*

Another common denial of the conflict between growth and the environment is the claim that because GDP is measured in value units, it has no physical impact on the environment. Although GDP is measured in value units, one must remember that a dollar's worth of gasoline is a physical quantity – recently about one-fourth of a gallon in EU countries. GDP is an aggregate of all such 'dollar's worth' quantities bought for final use, and is consequently a value-weighted index *of physical quantities*. GDP is certainly not perfectly correlated with resource throughput, but, for matter-dependent creatures like ourselves, the positive correlation is pretty high. Prospects for absolute 'decoupling' of resource throughput from GDP seem limited, even though much desired and discussed.[126]

Of course, opportunities for decoupling should be actively sought through technology.[127] However, the Jevons Paradox describes the human tendency to consume more of

[126] Victor (2008); see also Jackson (2009), Maxton and Randers (2016).

[127] UNEP's International Resource Panel has published two major reports on Decoupling: UNEP. 2011. Decoupling natural resource use and environmental impacts from economic growth. Lead authors: Marina Fischer-Kowalski and Mark Swilling. Nairobi. UNEP. 2014. Decoupling 2: Technologies, Infrastructures and Policy Options. Lead authors Ernst von Weizsäcker and Jacqueline Aloisi de Larderel. Nairobi.

what has become more efficient, outweighing a large part of the resource savings from efficiency and potentially leading to an even higher resource consumption rate in a growth economy. This is not to deny some real possibilities of 'Green Growth'.[128]

Ecological economists have distinguished *growth* (quantitative increase in size by accretion or assimilation of matter) from *development* (qualitative improvement in design, technology or ethical priorities) and have advocated *development without growth* – qualitative improvement without quantitative increase in resource throughput beyond an ecologically sustainable scale. In Sect. 1.1, the example of the LED was mentioned providing more light with a lot less energy. Hence, one could indeed say that there is no *necessary* conflict between qualitative development and the environment. But there is certainly a conflict between quantitative growth and the environment. GDP accounting mixes growth and development together, as well as costs and benefits. It is a number that confuses as much as it clarifies.

Economic logic tells us to invest in the limiting factor. Is it the number of chainsaws, fisher nets or sprinklers, or the size of forests, fish stock or freshwater that limits production? Economic logic has not changed, but the identity of the limiting factor has. The old economic policy of manufacturing more chainsaws, fish nets or sprinklers is now mostly uneconomic. Investments should shift to *natural capital*, which is now the limiting factor. In the case of fisheries, this means reducing the catch to allow populations to increase to their previous levels.

Traditional economists have reacted to this change of the limiting factor in two ways: first, by ignoring it – by continuing to believe that we live in the empty world; second, by claiming that human-made and natural capitals are substitutes. Even if natural capital is now scarcer than before, neoclassical economists claim this is not a problem because human-made capital is a 'near perfect' substitute for natural resources. In the real world, however, what they call 'production' is in fact transformation. Natural resources are transformed (not increased) by capital and labour into useful products and waste.

While improved technologies can certainly reduce wastage in the use of resources, as well as make recycling easier, it is hard to imagine how the fund of agents of transformation (capital or labour) can substitute for or replace the flow of that which is being transformed (natural resources). Can we produce a ten-pound cake with only one pound of ingredients, simply by using more cooks and ovens?

While a capital investment in sonar may help locate those remaining fish in the sea, it is hardly a viable substitute for there actually being more fish in existence. At the same time, the capital value of fishing boats, including their sonar, collapses as soon as the fish disappear. In the full world, certain types of growth thus become uneconomic.

[128] E.g. OECD (2011).

1.12.3 *The GDP Fallacy Again: Treating Costs as If They Are Benefits*

It is finally becoming broadly recognized that maximizing GDP, which was never intended to measure societal well-being, is not an appropriate goal for national policy. Although no single measure will satisfy all purposes, GDP gained enormous power to influence national and international economic policy because of the broad consensus surrounding its use over many years and countries. GDP interprets every expense as positive and does not distinguish welfare-enhancing activity from welfare-reducing activity. For example, an oil spill increases GDP because of the associated cost of clean-up and remediation, while it obviously detracts from overall well-being. Examples of other activities that increase GDP include natural disasters, most illnesses, crimes, accidents and divorce. GDP is more tightly correlated with throughput (cost) than with either measured welfare or self-evaluated happiness (benefit).

GDP also leaves out many components that enhance welfare but do not involve monetary transactions and therefore fall outside the market. For example, the act of picking vegetables from a garden and cooking them for family or friends is not included in GDP. Yet buying a similar meal in the frozen food aisle of the grocery store involves an exchange of money and is counted as a subsequent GDP increase. A parent staying home to raise a family or do volunteer work is not included in GDP and yet they make potentially key contributions to society's well-being.

In addition, GDP does not account for the distribution of income among individuals, which has considerable effect on individual and social well-being. GDP doesn't care whether a single individual or corporation receives all the income in a country, or whether it is equally distributed among the population. However, a dollar's worth of increased income to a poor person produces more additional welfare than a dollar's increased income to a rich person.

And yet, even with all the problems surrounding GDP, it is the most commonly used indicator of a country's overall performance. Using GDP as the yardstick, the global economy has grown eight- to tenfold since 1950, a vast increase in physical throughput.[129] The reason for the continued use of the GDP as a performance indicator is that it goes hand in hand with paid employment – and *this* carries an extremely high value in our societies.

Many alternative indicators have been proposed over the past few decades, as researchers have worked to consolidate economic, environmental and social elements into a common framework that would reflect genuine net progress (see Sect. 3.14).

[129] Higgs (2014, l.c., p. 34); Maddison (1995); World Bank annual figures, 1961–2015: World Bank. GDP Growth (annual %), http://data.worldbank.org/indicator/NY.GDP.MKTP.KD.ZG

Linking Chapters 1 and 2

Chapter 1 – like the whole book – was written 45 years after the publication of The Limits to Growth and 25 years after the 1992 'Earth Summit' of Rio de Janeiro. A new assessment of the impact of Rio is entitled After 25 years of trying, why aren't we environmentally sustainable yet?[130] Howes and his team reviewed 94 studies of how sustainability policies had failed across every continent. These included case studies from both developed and developing countries, and ranged in scope from international to local initiatives. The review concluded that since 1970 the biodiversity index has fallen by more than 50%, the human ecological footprint has risen to the point where 1.6 planets would be needed to provide resources sustainably, annual greenhouse emissions have almost doubled and the world has lost over 48% of tropical and subtropical forests.

The author found three recurring types of failure: economic, political and communication. Environmentally damaging activities are usually profitable; governments are unable or unwilling to implement effective policies; and communication fails to explain protection necessities to local communities, leading to massive opposition. And this happens around the world, North and South.

Regarding the way out, Dr. Howes suggests that governments provide financial incentives to switch to eco-efficient production and provide a viable transition pathway for industries that are doing the most damage. Business leaders from all sectors need to be convinced of both the seriousness of the declining state of the environment and that sustainable development is possible.

OK, that's a nice summary of the situation but far too harmless. Governments are not failing to communicate because they are stupid but because they would lose the next elections if they were honestly communicating. And businesses would soon be out of business if they were despising what is profitable. Nearly all actors are simply following what in the real world appears good for them.

One point is missing in Michael Howes' analysis: During these decades since 1970, world population has more than doubled, and per capita consumption rates as well. Humans are too many and too greedy, consuming everything available and failing to reflect on future generations' needs. In softer language UNDESA (United Nations Department of Economic and Social Affairs) says, 'Even if we succeeded in pushing our technological capabilities to the utmost, *without doing something else,* in a few decades we are likely to end up in a world that would offer reduced opportunities for our children and grandchildren to flourish'.[131]

Let us look at climate for an example. Countries agreed at the Paris COP 21 that CO_2 and other greenhouse gas emissions have to be reduced rapidly and significantly. That challenge is transported into the national debates, and there the first, almost automatic, response is that reducing emissions without reducing jobs and

[130] Michael Howes. 2017. After 25 years of trying, why aren't we environmentally sustainable yet? The conversation website (Australia).

[131] UNDESA (2012).

welfare will require a lot of additional money. Hence new stimuli for more economic growth are discussed first. *Without doing something else*, that would lead to more, not less, greenhouse gas emissions.

The four italicized words *without doing something else* can be understood as an admonition that everyone supporting a transformational agenda should heed. Keeping the grim facts of Chap. 1 in mind, it is clear that humanity must be prepared for a considerably more radical transformational agenda than simply investing in new technologies while supporting constant economic expansion and tolerating further population growth. The overall goal, so it seems, can no longer be just 'growth'. It should become truly 'sustainable development'.

To attain this, a serious transformational agenda must be defined and checked for consistency and for desirable purposes and outcomes. Humanity is faced with nothing less than establishing a new mind-set and a new philosophy, because the old growth philosophy is demonstrably wrong.

Two different *decoupling* tasks have to be pursued: decoupling the production of goods and services from unsustainable, wasteful or uncaring treatment of humans, nature and animals (do better), and decoupling the satisfaction of human needs from the imperative to deliver ever more economic output (do well).[132] The second task in effect means less GDP, which is anathema to all political parties, as is indicated in Sect. 1.12.3, where it was said that GDP goes hand in hand with paid employment, which nobody wantonly would risk to reduce.

For pursuing the transformational agenda of *sustainable* development, a new mind-set will be needed, which would favourably weigh the advantages of a sustainable world for future generations against high employment figures in our days. That, however, means a different political and civilizational philosophy for our era of the *full world*.

Chapter 2 of our book will therefore focus on philosophy, with the hope of arriving at some clues for an early sketch of a better philosophical framework. This search may lead to the desire for – if not the necessity of – a new 'Enlightenment'.

Of course, the Club of Rome is not alone aiming at a profound transition to a sustainable world: The United Nations Environment Programme (UNEP) in its fifth GEO Assessment[133] states, 'A transition to sustainability demands profound changes in understanding, interpretative frameworks and broader cultural values, just as it requires transformations in the practices, institutions and social structures that regulate and coordinate individual behaviour'. Similar intentions can be seen in the OECD's Innovation Strategy (2015 revision)[134] as well as in the Great Transition Network (GTN) initiated by Paul Raskin, director of the Tellus Institute in Boston. He imagines a global 'country', called *Earthland*, the place for a planetary civilization.[135]

When considering strategic options for overcoming the 'disarray' (Sect. 1.1) and the manifold features of non-sustainability (Sects. 1.2, 1.3, 1.4, 1.5, 1.6, 1.7, 1.8, and

---

[132] Göpel (2016), chiefly pages 20–21.

[133] UNEP GEO 5 Report, 2012, p. 447.

[134] OECD. 2015. The OECD Innovation Strategy. An Agenda for Policy action (2015 Revision). p. 6.

[135] Raskin (2016).

1.9), one should be aware of the potential dangers and opportunities of deep transformational change. But one of the most important steps for a proper assessment and mature judgement may be a better understanding of the 'philosophical crisis' of our time. Beyond the task of intellectual understanding, the philosophical analysis should help clarify where the potential partners stand in terms of a transition to values and mind-sets for true sustainability on Spaceship Earth.

References

Admati A, Hellwig M (2013) The bankers new clothes. Princeton University Press, Princeton

Arsenault C (2014) Top soil could be gone in 60 years if degradation continues, UN Official Warns. GREEN, Reuters, 5 Dec 2014

Bardi U (2014) Extracted. How the quest for mineral wealth is plundering the planet. A report to the Club of Rome. Chelsea Green Publishing, White River Junction

Bartlett B (2013) Financialization as a source of economic malaise. NY Times, 11 June 2013. https://economix.blogs.nytimes.com/2013/06/11/financialization-as-a-cause-of-economic-malaise/

Beder S (2006) Suiting themselves: how corporations drive the global agenda. Earthscan, London, p 42

Boulding K (1966) The economics of the coming spaceship earth. In: Jarrett H (ed) Environmental quality in a growing economy. Johns Hopkins University Press, Baltimore

Bower JL, Christensen CM (1995) Disruptive technologies: catching the wave. Harvard Business Review, Jan–Feb 1995

Braungart M, McDonough W (2002) Cradle to cradle: remaking the way we make things. North Point Press, New York

Brugmann J (2009) Welcome to the urban revolution. Penguin Books, London/New York

Chancel L, Piketty T (2015) Carbon and inequality: from Kyoto to Paris. Paris School of Economics, Paris

Corlett A (2016) Examining an elephant. Globalisation and the lower middle class of the rich world. Resolution Foundation, London

Crotty J (2009) Structural causes of the global financial crisis: a critical assessment of the 'new financial architecture'. Camb J Econ 33:563–580

Daly H (2005) Economics in a full world. Scientific American, September 2005, pp 100–107

Dempsey N et al (2016) Energy prices. House of Commons Briefing Paper 04153, London

Diamandis P, Kotler S (2012) Affluence. The future is better than you think. Free Press, New York

Diamandis P, Kotler S (2015) Bold: how to go big, create wealth and impact the world. Simon & Schuster, New York

Dugarova E, Gülasan N (2017) Six megatrends that could alter the course of sustainable development. The Guardian, 18 Apr 2017

Eggers D (2011) The circle. Knopf, New York

Ehrlich PR, Holdren JP (1971) Impact of population growth. Science 171(3977):1212–1217

FAO (2016) The state of world fisheries and aquaculture 2016. Rome

Georgescu-Roegen N (1971) The entropy law and the economic process. Harvard University Press, Cambridge, MA

Girardet H (1999) Creating sustainable cities, schumacher briefing 2. Green Books, Totnes

Göpel M (2016) The great mindshift. Springer, Berlin

Greenwood R, Scharfstein D (2013) The growth of finance. J Econ Perspectives 27:3–28

Higgs K (2014) Collision course: endless growth on a finite planet. MIT Press, Cambridge, MA

Hintemann R, Clausen J (2016) Green cloud? Current and future developments of energy consumption by data centers, networks and end-user devices. In: 4th international conference on ICT4S, Amsterdam, Aug 2016

Hoekstra AY (2013) The water footprint of modern consumer society. Routledge, London
Hsu PD, Lander ES, Zhang F (2014) Development and applications of CRISPR-Cas9 for genome engineering. Cell. 157(6):1262–1278. issn:1097-4172
Jackson T (2009) Prosperity without growth: economics for a finite planet. Earthscan, London, pp 67–71
Jackson T, Webster R (2016) Limits revisited. A review of the limits to growth debate. Creative Commons, London
Jamaldeen M (2016) The hidden billions. Oxfam, Melbourne
Khor M (2017) Shocks for developing countries from President Trump's first days. TWN News Service twnis@twnnews.net, 2 Feb 2017
Kolbert E (2014) The sixth extinction: an unnatural history. Henry Holt & Co, New York
Kristensen HM, Norris RS (2017) Status of world nuclear forces. Federation of American Scientists, Washington, DC
Kurzweil R (2006) The singularity is near. Gerald Duckworth, London
Lenzen M, Moran D, Kanemoto K, Foran B, Lobefaro L, Geschke A (2012) International trade drives biodiversity threats in developing nations. Nature 486:109–112
Liberti S (2013) Land grabbing: journeys in the new colonialism. Verso, London
Lietaer B, Arnsperger C, Goerner S, Brunnhuber S (2012) Money and sustainability: the missing link. Triarchy Press, Devon
Lübbert C, et al (2017) Environmental pollution with antimicrobial agents from bulk drug manufacturing industries in Hyderabad, South India. Infection, May 2017. doi:https://doi.org/10.1007/s15010-017-1007-2
Machovina B, Feeley KJ, Ripple WJ (2015) Biodiversity conservation: the key is reducing meat consumption. Sci Total Environ 536:419–431
Maddison A (1995) Monitoring the world economy, 1820–1992. OECD Development Centre, Paris
Martine G, Alves JE, Cavenaghi S (2013) Urbanisation and fertility decline: cashing in on structural change. IIED, London
Maxton G, Randers J (2016) Reinventing prosperity. Managing economic growth to reduce unemployment, inequality, and climate change. Greystone Books, Vancouver/Berkeley
McDonough W, Braungart M (2013) The upcycle. Beyond sustainability, designing for abundance. North Point Press, New York
McLean B, Nocera J (2010) All the devils are here, the hidden history of the financial crisis. Penguin, Portfolio
Meadows D, Meadows D, Randers J, 3rd Behrens W (1972) The limits to growth. Universe Books, New York
Monbiot G (2015) Grand promises of Paris climate deal undermined by squalid retrenchments. Guardian, 13 Dec 2015. https://www.theguardian.com/environment/georgemonbiot/2015/dec/12/paris-climate-deal-governments-fossil-fuels
National Academies of Sciences, Engineering, and Medicine (2016) Gene drives on the Horizon: advancing science, navigating uncertainty, and aligning research with public values. The National Academies Press, Washington, DC
Neslen A (2016) Leaked TTIP documents cast doubt on EU-US trade deal. The Guardian, 2 May 2016. https://www.theguardian.com/business/2016/may/01/leaked-ttip-documents-cast-doubt-on-eu-us-trade-deal
Obersteiner M, Walsh B, Frank S, Havlik P, Cantele M, Liu J, Palazzo A, Herrero M, Lu Y, Mosnier A, Valin H, Riahi K, Kraxner F, Fritz S, van Vuuren D (2016) Assessing the land resource-food price nexus of the Sustainable Development Goals. Sci Adv. https://doi.org/10.1126/sciadv.1501499
OECD (2011) Green growth and sustainable development. OECD, Paris
Oxford Poverty and Human Development Initiative (OPHI) (2017) Global multidimensional poverty index. Oxford
Patnaik P (1999) The real face of financial liberalisation. Frontline Magazine 16(4):13–26. http://www.frontline.in/static/html/fl1604/16041010.htm

Raskin P (2016) Journey to Earthland. The great transition to planetary civilization. Tellus Institute, Boston
Rifkin J (2011) The third industrial revolution. Palgrave Macmillan, London
Rockström J, Klum M (2012) The human quest: prospering within planetary boundaries. Princeton University Press, Princeton
Rockström J, Steffen W, Noone K et al (2009a) Planetary boundaries: exploring the safe operating space for humanity. Ecol Soc 14(2):1–32
Rockström J, Steffen W, Noone K et al (2009b) A safe operating space for humanity. Nature 461:472–475
Rome A (2015) Sustainability: the launch of spaceship earth. Nature 527:443–445
Romm J (2017) 2016 has crushed the record for hottest year. Think Progress, 01 Apr 2017
Sachs J, Schmidt-Traub G, Kroll C, Durand-Delacre D, Teksoz K (2016) SDG index and dashboards. Global report. Bertelsmann Stiftung and Sustainable Development Solutions Network, New York
Sankhe S, Vittal I, Dobbs R, Mohan A, Gulati A, Ablett J, Gupta S, Kim A, Paul S, Sanghvi A, Sethy G (2010) India's urban awakening. McKinsey, Boston/Bangalore
Sassen S (2009) Too big to save: the end of financial capitalism. Open Democracy, 1 April 2009. http://www.opendemocracy.net/article/too-big-to-save-the-end-of-financial-capitalism-0
Scharmer O (2009) Seven acupuncture points for shifting capitalism to create a regenerative ecosystem economy. MIT Presencing Institute. www.presencing.com
Schumpeter JA (1942) Capitalism, socialism and democracy. Routledge, London, p 139
Smil V (2011) Harvesting the biosphere: the human impact. Popul Dev Rev 37(4):613–636
Snyder T (2017) On tyranny. Twenty lessons from the twentieth century. Tim Duggan Books, New York
Steffen W, Crutzen PJ, McNeill JR (2007) The Anthropocene: are humans now overwhelming the great forces of nature? Ambio 36:614–621
Steffen W, Richardson K, Rockström J et al (2015) Planetary boundaries: guiding human development on a changing planet. Science 347(6223):736–747
Stout L (2012) The shareholder value myth. Berrett Koehler, San Francisco
Tibi B (2012) Islamism and Islam. Yale University Press, New Haven
Tsao J, Saunders HD et al (2010) Solid-state lighting: an energy-economics perspective. J Phys D Appl Phys 43:354001
Turner A (2016) Between debt and the devil: money, credit and fixing global finance. Princeton University Press, Princeton
Turner G, Alexander C (2014) Limits to growth was right. New research shows we're nearing collapse. The Guardian, 2 Sept 2014
UNDESA (2012) Back to our common future. Sustainable development in the 21st century (SD21) project. Summary for policymakers. United Nations (UN), New York, p iii
United Nations (2011) World urbanization prospects. UN, New York
United Nations Conference on Environment and Development (1992) Agenda 21. UNCED, New York. https://sustainabledevelopment.un.org/content/documents/Agenda21.pdf
van der Sluijs JP et al (2015) Conclusions of the Worldwide Integrated Assessment on the risks of neonicotinoids and fipronil to biodiversity and ecosystem functioning. Environ Sci Pollut Res 22(1):148–154
Victor P (2008) Managing without growth: slower by design, not disaster. Edward Elgar Publishers, Cheltenham, pp 54–58
Walter G, Weitzman ML (2015) Climate shock: the economic consequences of a Hotter Planet. Princeton University Press, Princeton
Waters CN, et al (2016) The Anthropocene is functionally and stratigraphically distinct from the Holocene. Science, 8 Jan 2016. http://science.sciencemag.org/content/351/6269/aad2622
Weiler RA, Demuynck K (2017) Food scarcity unavoidable by 2100? Impact of demography & climate change. Globethics.net. www.amazon.com/dp/1544617550/ref=sr_1_1
Williams ED, Ayres RU, Heller M (2002) The 1.7 kilogram microchip: energy and material use in the production of semiconductor devices. Environ Sci Technol 36(24):5504–5510

World Society for the Protection of Animals (WSPA) (2008) Eating our future. The environmental impact of industrial animal agriculture (Author: Michael Appleby). WSPA International, London
WTO (2010) Mexico etc. versus US: "Tuna-dolphin." http://www.wto.org/english/tratop_e/envir_e/edis04_e.htm
Zacharia F (2016) Populism on the march: why the west is in trouble. Foreign Affairs, Nov–Dec 2016
Blue Planet Prize Laureates (2012) Environment and development challenges: the imperative to act. Presented at UNEP, Nairobi, Feb 2012. Asahi Glass Foundation, Tokyo
Stockman D (2013) We're blind to the debt bubble interview with Paul Solman. PBS Newshour, 30 May 2013
Quattrociocchi W, Scala A, Sunstein CR (2016). Echo chambers on Facebook, 13 June 2016. Available at SSRN:https://ssrn.com/abstract=2795110
Fan R et al (2014) Anger is more influential than joy: Sentiment correlation in Weibo. doi:https://doi.org/10.1371/journal.pone.0110184
NCPA (2015) The 2008 housing crisis displaced more Americans than the 1930s Dust Bowl. National Center for Policy Analysis, 11 May 2015
Turner G (2008–09) A comparison of limits to growth with thirty years of reality, CSIRO working papers series
Sean Ó hÉigeartaigh (2017) Technological wild cards: existential risk and a changing humanity. Ethics, humanities, innovation, technology. https://www.bbvaopenmind.com/en/article/technological-wild-cards-existential-risk-and-a-changing-humanity/
HasselvergerL(2014)22factsaboutplasticpollution(And10thingsyoucandoaboutit).EcoWatch,7Aug 2014. http://ecowatch.com/2014/04/07/22-facts-plastic-pollution-10-things-can-do-about-it/
Civil Society Working Group on Gene Drives (2016) The case for a global moratorium on genetically-engineered gene drives. www.synbiowatch.org/gene-drives
GRAIN and La Via Campesina (2014) Hungry for land: small farmers feed the world with less than a quarter of all farmland. https://www.grain.org/article/entries/4929
UN (2009) Report of the Commission of Experts of the President of the United Nations General Assembly on Reforms of the International Monetary and Financial System. http://www.un.org/ga/econcrisissummit/docs/FinalReport_CoE.pdf
Vickery G (2012) Smarter and greener? Information Technology and the Environment: positive or negative impacts? International Institute for Sustainable Development (IISD), Oct 2012
Suhas Kumar (2015) Fundamental limits to Moore's Law. Cornell University. arXiv:1511.05956v1
European Union (2014) Critical raw materials. http://ec.europa.eu/growth/sectors/raw-materials/specific-interest/critical_en
Silicon Valley Toxics Coalition (2006) Toxic sweatshops. http://svtc.org/our-work/e-waste/
Climate Group for the Global eSustainability Initiative (2008) SMART 2020: enabling the low-carbon economy in the information age. http://www.smart2020.org/_assets/files/02_Smart2020Report.pdf
Frey CB, Osborne MA (2013) The future of employment: how susceptible are jobs to computerization? http://www.oxfordmartin.ox.ac.uk/downloads/academic/
World Economic Forum (2016) The future of jobs. Employment, skills and workforce strategy for the fourth industrial revolution. WEC
TIME (2017) Time events and promotion ad TIME Magazine, 27 Mar 2017, page 26; www.beyondsport.org

Chapter 2
C'mon! Don't Stick to Outdated Philosophies!

2.1 Laudato Sí: The Pope Raises His Voice

Pope Francis made significant headlines when he published an encyclical letter in June 2015, entitled *Laudato Sí*,[1] in which he squarely addressed the increasing destruction of our 'Common Home', the planet earth. He spoke out against toxic pollution, waste and the throwaway culture, as well as uncontrolled global warming, and the rapid destruction of biodiversity. He addressed, as the United Nations has also done, the growing economic gulf between the rich and the poor, and the seeming inability of nearly all countries to reduce this gap. He deplored the fact that many efforts to seek concrete solutions to the environmental crisis have proved to be ineffective, not only because of powerful opposition but also because of a more general lack of interest.[2]

The Pope went into considerable detail, describing the facts and dynamics of environmental destruction, before calling for a new attitude towards nature. In paragraph 76, he stated that 'Nature is usually seen as a system, which can be studied, understood and controlled, whereas creation can only be understood as a gift…' The message being that humanity needs to acquire an attitude of modesty and respect, rather than of arrogance and power.

Laudato Sí addresses the central problem of a widespread short-term economic logic which ignores the real cost of its long-term impact on nature and society:

> As long as production is increased, little concern is given to whether it is at the cost of future resources or the health of the environment; as long as the clearing of a forest increases production, no one calculates the losses entailed in the desertification of the land, the harm done to biodiversity or the increased pollution. In a word, businesses profit by calculating and paying only a fraction of the costs involved.[3]

---

[1] Pope Francis (2015).

[2] Ibid. paragraphs 14 and 20.

[3] l.c. paragraph 195.

© Springer Science+Business Media LLC 2018

E.U. von Weizsäcker, A. Wijkman, *Come On!*,
DOI 10.1007/978-1-4939-7419-1_2

Earlier in the document, the Pope wrote: 'The markets, which immediately benefit from sales, stimulate ever greater demand. An outsider looking at our world would be amazed at such behaviour, which at times appears self-destructive'. And later added, 'When human beings place themselves at the centre, they give absolute priority to immediate convenience and all else becomes relative'. Finally, he castigated the relativism of those who say 'Let us allow the invisible forces of the market to regulate the economy, and consider their impact on society and nature as collateral damage'.

The message of this historic encyclical is very clear: Humanity is on a suicidal trajectory, unless some strong, restraining rules are accepted that curtail the short-term utilitarian habits of our current economic paradigm. It could be wise to pay attention as well to the spiritual and religious dimensions of all civilizations that have counselled similar restraints. As the Pope put it, 'All of this shows the urgent need for us to move forward in a *bold cultural revolution*'.[4]

We have selected *Laudato Sí* as a means to begin this book's necessary discussion about environmental ethics and the religions of the world. However, 30 years earlier, the World Council of Churches (WCC), to which most of the Christian denominations (with the exception of Catholicism) belong, addressed very similar concerns. Beginning at the WCC's Sixth Assembly in Vancouver, 1983, the churches present, feeling the dangers of conflict, including a third world war, called for the convening of an all-Christianity 'Peace Council'. Discussions on the causes of armed conflicts led to a decision to add justice and 'the integrity of creation' to this agenda. Based on the general mandate from Vancouver, discussions continued, ultimately leading to a convocation on *Justice, Peace and the Integrity of Creation* in Seoul, Korea, in March 1990. Ten 'Affirmations' were approved, covering the three pillars of justice, peace and the integrity of creation. The seventh of these Affirmations addressed the nexus between peace, justice and the environment, overtly recognizing the self-renewing, sustainable character of natural ecosystems, that is, of God's creation. The convocation's language and its credible foundation in both Christian tradition and the Bible demonstrate a strong resemblance with the later *Laudato Sí*.

Less recognized in Western circles but of similar clarity in language is the 2015 Islamic Declaration on Global Climate Change, which states: 'The epoch in which we live has increasingly been described in geological terms as the Anthropocene, or "Age of Humans". Our species, though selected to be a caretaker or steward (khalifah) on the earth, has been the cause of such corruption and devastation on it that we are in danger ending life as we know it on our planet. This current rate of climate change cannot be sustained, and the earth's fine equilibrium (m+z n) may soon be lost. As we humans are woven into the fabric of the natural world, its gifts are for us to savour'.[5]

That declaration was the outcome of a year-long worldwide consultation process started by the Islamic Foundation for Ecology and Environmental Sciences

[4] l.c. paragraph 114; italics ours.

[5] http://islamicclimatedeclaration.org/islamic-declaration-on-global-climate-change/

(IFEES/EcoIslam). It was supported by Islamic Relief Worldwide before being discussed through the Climate Action Network and the Forum on Religion and Ecology. While it was not issued by internationally visible lead figures of Islam, it stands for a wide network of Muslim-led initiatives and thinkers. One quote may suffice to illustrate the language chosen: 'In reminding the richer nations to shoulder their proportion of accountability for creating the greater volume of this problem, it behoves each single one of us to play our part in returning the Earth to some semblance of balance'.[6]

Islam's capacity to bring the Qur'an into a creative symbiosis with the rational sciences and with other non-religious features of human society has its roots in early medieval thinking. Avicenna/Ibn Sina (ca 980–1037), an eminent Islamic doctor and scientist from Bukhara and later Persia, quoted the Qur'an to refute astrology because it was not fact-based; his rational and factual approach made him one of the first serious astronomers of the world, and his science-based medicine became standard reading for centuries for all doctors in the Western world. Averroës/Ibn Rušd (1126–1198), living mostly in what is now Spain, following both Avicenna and Aristotle, also became an eminent doctor and scientist and is often quoted as the towering figure of an early Islamic Enlightenment. Unfortunately, radical Islamic schools of today tend to ignore or fight this approach of a symbiosis between the Islamic faith and science.

The late Judge Christopher Gregory Weeramantry,[7] former vice president of the International Court of Justice, has written a book summarizing key texts on humanity's responsibilities towards nature, other forms of life and all future generations, as found in the scriptures of five major world religions. In his introduction, the Sri Lankan judge writes that it is surely paradoxical that the latest generation in humanity's 150,000 years of existence today ignores the wisdom of those 150 millennia, which are enshrined in the core teachings of the world's great religions. Weeramantry raises concerns that two tendencies, the secularization of the state and the emergence of international law, a judicial form that has become entirely independent from ethical teachings common to all the world's greatest religions, have taken modern society too far from its key moral beliefs. He suggests integrating the principles of the major religions into international law, in order to properly address the current crises facing humanity.

And yet, some religions, including Judaism and Christianity, do contain teaching that justify human's dominance and can lead to human negligence towards nature. The famous *dominium terrae* story (Genesis 1: 26–28) is often used as an example of this. It reads[26]: 'Then God said, "Let us make mankind in our image, in our likeness, so that they may rule over the fish in the sea and the birds in the sky, over the livestock and all the wild animals, and over all the creatures that move along the ground".[27] So God created mankind in his own image, in the image of God he created them; male and female he created them.[28] God blessed them and said to them, "Be fruitful and increase in number; fill the earth and subdue it"'.

[6] http://www.ifees.org.uk/declaration/#about

[7] Weeramantry (2009).

Going back to the origins of the major religions of the world, it should be recognized that they all emerged at times when nature seemed robust and endless, and the relatively few humans were threatened by hunger, wild animals, unknown diseases and neighbouring tribes. That situation was characteristic of Herman Daly's concept of the 'empty world' discussed in Chap. 1. Even so, wise elders in the communities of that so-called *empty world* understood the need for long-term thinking, including provisions for storing food for winter or against time of famine, planning complex expeditions and creating a legal framework for the orderly functioning of social communities. The elders might generally think of divine power as inaccessible to humanity, but as also giving guidance for everyday life. Divine power often represented the long-term perspective, including eternity.

Early stories of gods and goddesses were often linked to the fortune of warriors as in the ancient Greek Iliad and Odyssey sagas. This tradition of various gods helping their 'chosen people' to survive and to triumph their adversaries continues today. 'Holy' wars including 'colonization' wars were fought throughout the ages. In our days, the once much more narrowly used Islamic *jihad* is seen by its warriors as just and inevitable when infidels insult God and his followers. The historian Philippe Buc, and in a different manner Karen Armstrong, discusses the long propensity for violent aggression present since the early days of Christianity,[8] echoing of course similar traditions in Judaism. But Armstrong argues that there is no inherent violence in religions themselves. The Club of Rome does not support aggressive, warlike religious doctrines wherever they may be found, but does feel much can be achieved by paying more attention to the many pan-religious directives encouraging believers to take good care of creation and the Common Home. It also needs to be recognized that the mandate to 'be fruitful and increase in number; fill the earth and subdue it', common to all three Abrahamic religions, Judaism, Christianity and Islam, can no longer apply under the new conditions of our full world.

2.2 Change the Story, Change the Future

One new approach to the debate about the role of religion in the environmental and social crises of today was introduced by David Korten in a recent Report to the Club of Rome.[9] He points out that historically it was mostly the three, closely related monotheistic (or Abrahamic) religions, Judaism, Christianity and Islam, that have been able to persist and especially to greatly expand over the millennia. Korten characterizes all three as using the same story – that of a *Distant Patriarch* ruling over humanity and His creation, nature. This often-comforting story of the Distant Patriarch comes with problematic side effects, however, which include the almost

[8] Philippe (2015) and Armstron (2014).

[9] Korten (2014).

permanent use of military power, the build-up of political/religious elites, the oppression of women, the prosecution of intellectuals and an inherent rigidity of doctrine. Such features almost inescapably then give rise to opposing movements towards freedom and *Enlightenment*, typically in opposition to the current church hierarchy, but mostly faithful to the original religious wisdom.

The European version of one major reaction to the Distant Patriarch, the Enlightenment of the seventeenth and eighteenth centuries, prompted the rise of science, technology and a subsequent adoration of technological 'miracles'. This, Korten says, resulted in a brand-new 'story': the *Grand Machine* cosmology. 'The contributions of science to human advancement and well-being, knowledge, and technology give this cosmology considerable authority and respect', writes Korten.[10] But it also involved conferring a 'sacred' character onto money, eventually resulting in a world ruled by 'money seeking robots'.[11]

In order to avoid the rigidity of the Distant Patriarch cosmology and the destructive 'sacred money' paradigm, Korten proposes a new social narrative and a new cosmology called the *Sacred Life and Living Earth* story, to which he devotes the rest of his book. He uses self-governing local communities, the 2010 Cochabamba Declaration on the Rights of Mother Earth, and emerging movements worldwide towards a *Living Economy* as examples of how the earth and all the life on it can be preserved by wide adherence to a different kind of story.

This report does not claim to have the right answers to all these questions. But it should be stressed that addressing today's challenges, as the Pope, the WCC, the IFEES and Korten and other authors are arguing, will necessarily involve a *spiritual dimension*, a moral vantage point. To address the daunting issues before us, it is simply not acceptable that selfishness and greed continue to enjoy positive social connotations as supposed drivers of progress. Progress can flourish just as well in a civilization that fosters solidarity, humility and respect for Mother Earth and for future generations.

2.3 1991: 'The First Global Revolution'

In 1991, Alexander King and Bertrand Schneider co-authored a powerful book, which they called *The First Global Revolution*.[12] In this they distinguished the *problematique*, which had been outlined in *The Limits to Growth*, from the *resolutique*, a new term in English or French indicating that they were now addressing the actual ways and means of successfully dealing with those problems.

The leading organ of the Club of Rome at the time was called the *Council of the Club of Rome*. And King and Schneider obtained the agreement – however

[10] Ibid., p. 40.

[11] Ibid. pp. 25–27 and 87–97.

[12] King and Schneider (1991).

Fig. 2.1 *The first global revolution. A report by the Council of the Club of Rome.* Lead Author Alexander King, president of the Club of Rome 1984–1990, president emeritus thereafter (Book cover: own photo; picture A. King: Courtesy Alex King's family)

hesitant – by the Council to call the new book *A Report by the Council of the Club of Rome* (Fig. 2.1).

Significantly, *The First Global Revolution* saw the end of the Cold War as an enormous opportunity for humanity to change course and unify, through the identification of a new 'common enemy'. The new enemy was the *problematique* of environmental degradation and global warming, poverty, military overspending and the scarcity of resources, including energy and water. The governments of the world had to cooperate in confronting these dragons. *Good governance* was one of the key expressions used in the book, a major component of the *resolutique* that would consist of international campaigns to overcome hunger, water scarcity, militarization and so on. *The First Global Revolution* served as a push for the Agenda 21, adopted in 1992 at the UN Earth Summit of Rio de Janeiro, which was updated in the 2015 Sustainable Development Goals.

However, the visionary programme Agenda 21's never really materialized. A purist pro-market ideology got the upper hand (see the two next sections) and was fervently opposing the idea of spending hundreds of billions of tax dollars for financing Agenda 21. In the end, the Club of Rome's *resolutique* – which may have helped avert some of the crises we face today – shared the fate of Agenda 21 and fell into oblivion.

2.4 Capitalism Got Arrogant

Historians know that the new market-based mind-set consolidated its influence after the end of the Cold War. To understand the philosophical nature of that new mind-set and its momentum, it may be useful to comprehend the power array before 1989, beginning with the setting immediately after World War II (WWII).

It was clear in 1945 that a repetition of the disasters of another world war must be avoided under all circumstances. The United Nations was founded with that specific purpose expressed at the start of its Charter: *To maintain international peace and security, and to that end: to take effective collective measures for the prevention and removal of threats to the peace.*

Soon after the creation of the United Nations, however, a deep divide emerged among the victorious powers. On one side was the Soviet Union, having lost more than 20 million lives. On the other side were the Western democracies, led by the United States and joined by Britain and France. In the years following WWII, the Soviet Union occupied or annexed most of the countries of Eastern Europe, forcing them to adopt Soviet communism as their political model. When Czechoslovakia also succumbed to Soviet dominance, and Mao Zedong established a communist government in China soon after, the West began to panic, and the *Cold War* began.

The Soviet credo was that capitalism was impoverishing the masses and therefore, if necessary, had to be defeated by military power. Realizing that there was some dangerous magnetism to the Soviet claim, the West began eagerly trying to demonstrate that a free and democratic market economy could in effect be more attractive for the masses, if it were also taking care of the needs of poor and disadvantaged people. That was an important motive for the formation of what came to be known as the welfare state or the *social* market economy.

All Western countries developed redistributive taxation systems, with marginal income tax rates as high as 90% for the rich, even in the United States. The generous US Marshall Plan after the war supported a stunning resurrection of Europe and of Japan. As divided countries, Korea and Germany willy-nilly became experimental playgrounds for testing whether communism or a free and social market economy worked better for the poor as well as for the rich. The Western strategy succeeded. Economic growth 'lifted all boats', and poverty declined.

After 40 years, the ideological 'war' ended with the victory of the West. Communism collapsed (except, oddly, in North Korea), and Francis Fukuyama, in 1989, declared the 'end of history'.[13] What was meant by this expression was the widely held belief that the free, democratic market economy was not only victorious over one specific adversary, Soviet communism, but was the best possible system, period.

The trouble was that – in full accordance with the basic tenet of *competition* in the market theory – the absence of a rival made the victorious party arrogant. With political roots that dated back to Pinochet's Chile (1973), Thatcher's Britain (1979)

[13] Francis Fukuyama. 1989 and 1992, l.c.

and Reagan's America (1981), after 1989 a *radical* market philosophy became the new mind-set of the entire world. Liberalization, deregulation and privatization[14] became the uncontested melody of the political agenda of the 1990s, legally culminating in the creation of the World Trade Organization (1994) – after the completion of the Uruguay Round of the GATT that radically strengthened the muscles of markets and correspondingly weakened those of nation states.

The phenomenon of 'globalization' meant essentially that small and medium states had to surrender much of their governing powers to the markets. In the EU, a *voluntary* reduction of national sovereignty was achieved with the creation in 1957 of the European Economic Community. But the ascendancy of 'free market' thinking, especially after 1989, eroded sovereignty everywhere and, in the EU and all WTO signatory states, further weakened *public* powers.

Weakened public powers also meant that poor and disadvantaged people could no longer truly rely on the state for help. State revenues slipped downwards, as tariffs were removed and a competition set in among states to attract investors through reduced taxation. In the end, a new phenomenon emerged, namely, *failed states*, more or less in parallel with the rise of globalization.

As German economist Hans Werner Sinn put it in 2003, the old *systems competition* (capitalism vs. communism) took place within closed borders. *Globalization* has brought about a new type of systems competition that is driven by the mobility of factors of production, especially capital, across borders. That new systems competition, he said prophetically, 'will likely imply the erosion of the European welfare state, and induce a race to the bottom in the sense that capital will not even pay for the infrastructure it uses and will erode national regulatory systems. In general, it will suffer from the same type of market failure, which induced the respective government activity in the first place'.[15]

One completely unintended side effect of this development is that average people, notably in younger generations, have begun to doubt if it is still worthwhile to vote. The world is therefore confronted with an increasing *crisis of democracy*. Re-strengthening democracy will require us to re-establish a fair balance between markets (mostly representing the private well-being of the high achievers) and the state (representing public goods and the interests of those who tend to be market losers or have no stake in them whatever). Markets also have a strong tendency to take a short-term view, while public interest must always include a long-term perspective.

The Club of Rome sees itself as an advocate of democracy, of long-term thinking, of the young and the unborn generations and of nature, which has no voice in capitalism and in the political debates among humans.

The agenda of rebalancing the public with the private good may take a whole generation, some 30 years. In this agenda, the intellectual and political weaknesses and strengths contained within markets and states should be identified. Neither

[14] See, e.g. von Weizsäcker et al. (2005).

[15] Sinn (2003).

purist market ideology nor pure state dominance will be acceptable, but very considerable synergies between the two can emerge from a suitable and balanced division of labour. This will only be possible with an engaged citizenry, able to hold leaders of both the public and private sectors to account.

2.5 The Failure of the Market Doctrine

In the preceding section, it was argued that capitalism became arrogant once it was the only game in town. The period since 1989 was also the period during which the *finance* sector extended its domination over the worldwide economy. Prior to the collapse of communism, the most powerful private sector actors included the big manufacturing, mining and service corporations, as well as large banks and insurance companies. But by 2011, 45 of the top 50 transnational corporations were not producing goods or services at all, but were banks or insurance companies.[16] Gradually, financial corporations became major shareholders and the real stewards of the productive companies. *Shareholder value* and *returns on investment* (RoI) became favourite expressions in the business world. Big investors would give the CEOs of manufacturing or service companies instructions on the minimum RoI to be achieved. Typically, the goal everything was intended to serve was the short-term quarterly report, a matter of weeks.

One could perhaps tolerate such an inversion of power, from the state to the corporations to the owners of capital, if the whole system had remotely lived up to its claims of 'lifting all boats' and creating universal prosperity. But, as Graeme Maxton and Jørgen Randers[17] write, today's capitalism is actually making many things worse, both for the environment and for people.

Maxton and Randers name a great number of dangerous failures: climate change, pollution (especially of the oceans), resource depletion and biodiversity loss; poverty, inequality and social friction; and unemployment, especially for the young. Civil, religious and territorial wars are also increasing along with terrorism, partly as a result of these endemic problems. These multiple crises are driving mass migrations of refugees. Finally, geopolitical friction is also rising. Many of the wars in recent decades have implicitly been about access to the resources needed to fuel economic growth, notably oil and water. Such wars, ironically, both push social and environmental problems further down the agenda and exacerbate them. Most of the migrants across Asia, as well as those from the Middle East and North Africa, are leaving areas riven by conflict and ravaged by poverty and resource destruction.

Maxton and Randers see this daunting package of problems as the result of the current economic system. All of them have the same basic cause: 'the desire for endless consumption growth without due concern for the impacts on the environment

---

[16] Vitali et al. (2011).

[17] Maxton and Randers (2016).

and inequality'. The capitalist incentive structure rewards cost cutting and short-term profits. It also 'generates constantly rising labour productivity and, unless it also creates a sufficient number of new jobs, this increases long-term joblessness'.

Extreme free-market thinking is at the root of the damage humanity is inflicting on the planet. 'The current *economic system requires a steady rise in the throughput of raw materials...* and, according to such thinking, the oceans, forest ecosystems, and polar ice have no economic value beyond the resources they can provide – the cost of the damage done to them tends to be completely ignored'.[18]

Such views are shared by a broad range of analysts and experts. Jean Ziegler writes that our current problems and disasters are chiefly caused by unbridled capitalism.[19] Even mainstream economists have come to the conclusion that markets do nothing to reduce inequality among people – quite the reverse. One of the most prominent economists who shares this view is Joseph Stiglitz, a new member of the Club of Rome.[20] And Thomas Piketty has published a deep historic analysis of the functioning of capitalism demonstrating that poverty eradication has simply never happened under the rule of capital.[21]

Similarly, Anders Wijkman and Johan Rockström, in their Report to the Club of Rome – *Bankrupting Nature* – have illustrated the fact that the destruction of nature and the emergence of financial collapses have followed essentially the same logic of greed, impatience and short-termism.[22]

A large group of economists and other thinkers who have joined in the 'Great Transition Network' (GTN) share online discussions about the state of the world: what needs to change and what can be done. The initiator and head of the group, Paul Raskin founded the Tellus Institute in Boston, which sponsors the Network.[23] The GTN including many members of the Club of Rome is a big tent, welcoming many positions, but most participants reject extreme free-market economics.

The rapid advance of the digital economy could make job destruction worse, as argued by Brynjolfsson and McAfee in their seminal book *The Second Machine Age*[24]: 'The faster growth is the more companies tend to invest in automation and robotisation'.

The people who promoted the extreme market model in modern times met in 1947 in Mont Pèlerin above Vevey, a small place in Switzerland. The meeting was convened by Friedrich von Hayek, a respected economist who later won the Nobel Prize for Economics. The meeting included Milton Friedman along with other well-known economists and fiscal experts. They were united at the time by a fear of

[18] Ibid.

[19] Ziegler (2014).

[20] Stiglitz (2012).

[21] Piketty (2014).

[22] Wijkman and Rockström (2012).

[23] Writings: Raskin (2014). The initial publication was Raskin et al. (2002). For contacts: gtnetwork@greattransition.org

[24] Brynjolfsson and McAfee (2014).

Fig. 2.2 Snapshot with Milton Friedman (*centre*) during the first meeting of the Mont Pèlerin Society (Source: www.montpelerin.org)

government expansion, especially of the welfare state, which they saw as 'dangerous'. They saw trade unions as 'dangerous', too – but the market as something close to divine. The group gave itself the name of the Mont Pèlerin Society (MPS). Hayek claimed that the intention was to create a venue for free and independent exchange of thought, not intervention in politics. Ralph Harris, however, a British economist who joined the MPS in 1960, acknowledged that the group's actual goal was to 'launch an intellectual crusade aimed at reversing the tide of post-war collectivism' (Fig. 2.2).[25]

It was not until the late 1970s, during the 'stagflation' crisis, that MPS ideas became dominant in conservative academic and political circles, and this form of neo-liberal economic thought began to succeed politically. Stagflation meant the simultaneous rise of inflation and stagnation. MPS representatives like Milton Friedman now had a chance to blame Keynesianism as the cause of this problem and to recommend a radical reduction of state intervention. Once Margaret Thatcher came to power in Britain and Ronald Reagan in the United States, they quickly began to implement the Mont Pèlerin ideas. In Reagan's teams of economic advisors, 22 were members of the MPS.

After a bumpy start, Reagan and Thatcher were successful in fostering economic growth and creating jobs. The disciples of the MPS conveniently claimed that the economic upswing was due to their new policies of tax cuts and reduced state intervention. However, the real reasons, one can argue, had little to do with their neo-liberal agenda. Probably more important was the surprising fact that, less than 10 years after the 'oil shock' of 1973, petrol and gas prices began to tumble (Fig. 2.3), eventually reaching levels (in constant dollars) close to the pre-shock prices.

This unexpected reversal of the high prices of 1973–1981 was caused by aggressive exploration and exploitation of new sources of oil and gas, demonstrating that the world had not yet exhausted its supply of cheap oil resources. Post-1982 cheap

[25] Harris (1997). See also Higgs (2014), Chapters 6, 10 and 11.

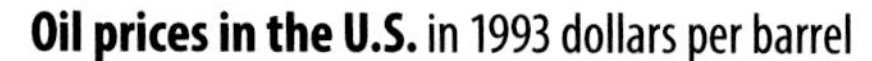

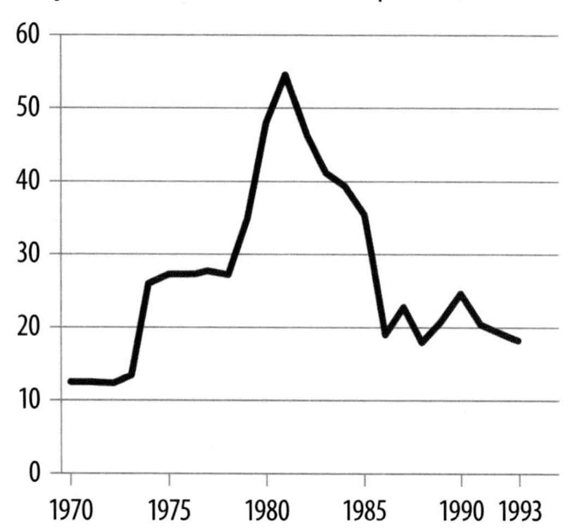

Fig. 2.3 Oil prices peaked in 1981 and tumbled throughout the first period of Ronald Reagan's presidency (Source: isgs.illinois.edu)

oil prices massively reduced inflation and transport costs, encouraging industries in the OECD and several emerging economies to greatly expand. Low oil prices also boosted new housing investments, chiefly in the United States, assisting the related supply industries.

Tragically, the 1980s were disastrous for many developing countries that had borrowed heavily to invest in resource extraction, counting on steadily rising prices; when resource prices collapsed, and the United States hiked interest rates, those countries found themselves in an intractable *debt crisis.*

By the 1980s, neo-liberal thinking had become a mainstream view in US academic circles and a broadly accepted alternative to the European social market economy. However, as long as the Soviet Union was seen as a threat, the West still needed to prove that the political system of a market economy served the poor better than socialism, so the really *extreme* form of free-market thinking remained a minority view in most countries outside the English-speaking world.

Today the collapse of communism is history, and free-market fundamentalism, in combination with the steeply rising influence of the finance sector, is our new reality. An understanding of the negative effects of extreme market thinking and practice has not reached average citizens. Nonetheless, increasing numbers of voices[26] can be heard, even from inside institutions like the International Monetary Fund (IMF), recognizing that neo-liberalism has been 'oversold' and that fiscal austerity policies can actually be detrimental to a country's economy.

The destructive effects of free trade agreements (in combination with ongoing automation and robotization) on the once-secure workers of America's manufacturing industries were exploited very effectively by Donald Trump in his presidential campaign. Once unionized supporters of the Democratic Party, many 'rust belt' workers voted for Trump. Given his win, it's ironic that both political parties in the

[26] Ostry et al. (2016).

United States have driven the 'free trade' agenda for 70 years, ever since Bretton Woods and GATT, and that both have continued to push for ever-intensified trade liberalization, largely on behalf of the corporations based in the country.

Whether the Trump administration will categorically diverge from this agenda remains to be seen. This may hinge on whether Americans have indeed lost faith that free trade is in their interest, which they were persuaded to believe in the past.

Yet another shortcoming of the liberalization of capital flow, an essential element in free trade, can be seen in the mechanisms large corporations have established to avoid taxation. In 2016 it became public knowledge that Panama, only one of many countries involved in such revelations over the past two decades,[27] has massively facilitated corporations and rich individuals in hiding their incomes from national tax authorities. Estimates of the amounts that are stashed in tax havens like the British Virgin Islands, the Cayman Islands and many other secrecy jurisdictions worldwide vary from US$21 trillion to US$32 trillion.[28] These arrangements serve to enhance the power of the finance sector and transfer ever more wealth to those who are already, in many cases, obscenely wealthy.[29]

So far, despite protectionist challenges to compulsory free trade, the free-market doctrine still prevails, and the transnational financiers it has favoured still control much of the global economy. However, as far as the planet and the vast majority of its people are concerned, this doctrine has failed us all.

2.6 Philosophical Errors of the Market Doctrine

To continue to criticize the failures of current market doctrine would only repeat what has already been demonstrated by many other authors. Instead it may be useful to look at the history and validity of some of the basic tenets of economics. Three major tenets in terms of relevance deserve further investigation and clarification:

- Adam Smith's concept of the *invisible hand* and the related conviction chiefly from the Chicago School of Economics that markets by definition are superior to states or to lawmakers in finding the optimum path of development
- David Ricardo's discovery of *comparative advantages*, which in theory makes trade a win-win operation for both sides of the transaction
- Charles Darwin's conceptual edifice, which has been *wrongly interpreted* as postulating that competition, the fiercer the better, leads to continued progress and evolution

All three tenets contain valid points, but all three have to be better understood and put into an historical perspective.

[27] Obermayer and Obermaier (2016).

[28] Henry (2012)

[29] Obermayer and Obermaier. l.c.

Fig. 2.4 This picture from the 2001 seasons greetings card by the Adam Smith Institute sees the institute's name patron as the saviour coming out his box preaching free markets to the world – and the world smiles gratefully (Courtesy Adam Smith Institute)

2.6.1 Adam Smith, Prophet, Moralist, Enlightener

Adam Smith is often seen as the early prophet of free markets. A funny document of the view is the 2001 Seasons Card by the Adam Smith Institute – Fig. 2.4.

This is clearly a caricature. In reality, Adam Smith was a complex social ethicist. Together with David Hume and John Locke, he represented the Enlightenment in Britain. The principle developed by Smith that has had the most impact was that the 'invisible hand' (a divine notion at his time)[30] would turn the pursuit of self-interest into common benefit, because economic self-interest in good-quality work would enhance the benefits of overall production.

However, one condition for Smith's logic was that the geographical reach of the law and of morality was identical to the geographical reach of the market, of that *invisible hand*. This fact, unquestioned in the eighteenth century, established a healthy balance between markets and the law. Even if markets have that admirable capacity to 'discover' the right prices and innovation opportunities, in Adam Smith's world they would still be restricted by firm legal or moral rules. Moreover, in Smith's time, markets were small, and trade was between rather small partners.

By contrast, trade in our day is dominated by large global corporations. Today's markets have *the world* as their geographical reach, while moral conventions and legal restrictions typically apply only to a nation or a specific culture. This leads to the phenomenon of economic globalization where markets, chiefly capital markets, can induce lawmakers to adjust the law to please investors and shareholders. 'It seems clear that the political manageability of democratic capitalism has sharply declined in recent years', says Wolfgang Streek.[31] Adam Smith's tacit assumption of a healthy balance between markets and the law is thereby ignored at its very core.

---

[30] Smith first introduced the notion in his The Theory of Moral Sentiments (Edinburgh, 1759).

[31] Wolfgang Streek (2011) The crisis of democratic capitalism. Europe Solidaire Sans Frontières.

An updated economic theory has to include mechanisms for re-establishing that healthy balance and also for providing a distinct place for moral codes. Political action should try to enlarge, rather than restrict, the reach of law, for example, by legally binding international conventions, where laws could increase transportation costs by removing subsidies thus giving economic advantages to local value creation. Such actions bring the reach of the law closer to the reach of the markets, that is, to Adam Smith's logic.

2.6.2 David Ricardo, Capital Mobility and Comparative Versus Absolute Advantage

It is often said that in a globalized economy countries and companies have no choice but to compete in the global quest for growth. This is not true. Globalization of the kind that evolved in the 1990s and thereafter was a *policy choice by our elites*, not an imposed necessity. On this issue there has so far been broad agreement between the centre-right and the centre-left, meaning there has been very little questioning of the basic postulates.

The Bretton Woods system was a major achievement aimed at avoiding the monetary chaos and competitive devaluations that caused the Great Depression of the 1930s. The currency stability it created fostered international trade for mutual advantage among countries. Free capital mobility and global integration were not part of the original deal, however, even though the United States pressed for an International Trade Organisation (ITO) from the start and its trade representative negotiated a general agreement on tariffs and trade (GATT) with 27 countries in 1947. GATT was progressively expanded over the years and by 1995, when it was restructured as the WTO, 108 countries had joined and tariffs had been slashed by 75%. Cross-border financial flows began to increase in the 1970s and exploded since the 1980s with the deregulation of domestic banking in many countries and the beginnings of electronic trading. After 1995, the WTO began pushing for unrestricted and compulsory *capital mobility*, assisted by the comprehensive deregulation of banking in the United States in 1999 (see Sects. 1.9 and 2.5).

Globalization is the engineered integration of many formerly relatively independent national economies into a single tightly bound global economy organized around *absolute advantage*, not *comparative advantage*. Comparative advantage occurs when one country can produce a good or service at a lower opportunity cost, meaning it can produce a good relatively cheaper than another. The theory of comparative advantage states that if countries specialize in producing goods where they have such lower costs, there will be an increase in general economic welfare. *Absolute advantage,* however, means the ability to produce goods or services using fewer actual inputs. So, for example, an industry growing tomatoes in Mexico, where sun is abundant, has an absolute advantage over tomato-growing in greenhouses in Canada.

Once a country has been sold on free trade and free capital mobility, it has effectively been integrated into the global economy and is no longer free to decide what to trade and what not. Yet all of the theorems in economics about the gains from trade assume that trade is voluntary. How can trade be voluntary if you are so specialized as to be no longer free not to trade? Countries can no longer account for social and environmental costs and internalize them in their prices unless all other countries do so, and to the same degree.[32]

To integrate the global omelette you must disintegrate the national eggs. While nations have many sins to atone for, they remain the main locus of community and policymaking authority. It will not do to disintegrate them in the name of abstract 'globalism', even though some global federation of national communities can be helpful. But when nations disintegrate, there will be nothing left to federate in the interest of legitimately global purposes. 'Globalization' (national disintegration) was an actively pursued policy, not an inertial force of nature, although greatly facilitated by technology. It can be undone, to an extent, as is the intention of the US government since 2017.

The IMF preaches free trade based on *comparative* advantage, and has done so for a long time. More recently, the WTO-WB-IMF have started preaching the gospel of globalization, which, in addition to free trade, means free capital mobility internationally, and increasingly free migration. The classical comparative advantage argument by David Ricardo, however, explicitly assumes international *immobility* of capital (and labour). Capitalists are interested in maximizing absolute profits and therefore generally seek to reduce absolute costs. If capital is mobile between nations, it will move to the nation with lowest absolute costs.

Only if capital is internationally immobile will capitalists have any reason to compare internal cost ratios of countries, and choose to specialize in the domestic products having the lowest relative cost compared to other nations, and to trade that good (in which they have a *comparative advantage*) for other goods. In other words, comparative advantage is a second-best policy that capitalists will follow only when the first-best policy of following absolute advantage is blocked by international capital immobility. This is straight out of Ricardo,[33] but is a facet of his thought that is far too often ignored. It is therefore very puzzling to see the IMF and some trade theorists advocating free trade based on comparative advantage, and at the same time advocating free capital mobility – as if the latter were merely an extension of the comparative advantage argument rather than the denial of its main premise.[34]

There are of course global gains from specialization and trade based on absolute advantage, just as there are from comparative advantage. In theory global gains from absolute advantage should be greater because specialization is not constrained by international capital immobility. However, under absolute advantage some countries gain and others lose from trade, whereas under comparative advantage,

[32] To be sure, conventional economists are less impressed by such "changes". A perfect example is Baldwin (2010); his emphasis on technological differences is worth deeper consideration.

[33] Ricardo (1951).

[34] For further discussion, see Daly and Farley (2004).

although some gain more than others, there are no losers. It is this mutual benefit guarantee that has been the main strength of the free trade policy based on comparative advantage. Theoretically, the global gains under absolute advantage could be redistributed from winners to compensate losers – but this would no longer be 'free trade' and most certainly is not being done!

By contrast, neo-liberal economists dominating the WTO, the IMF and other organizations wave their hands when confronted with this contradiction. They suggest that you might be a protectionist isolationist xenophobe, and then change the subject. The WTO-WB-IMF contradicts itself in service to the interests of transnational corporations and their policy of off-shoring production in the pursuit of cheap labour while falsely calling it 'free trade'.

International capital mobility, coupled with free trade, allows corporations to escape from national regulation in the public interest, playing one nation off against another. Since there is no global government, they are in effect uncontrolled. The nearest thing we have to a global government (WTO-WB-IMF) has shown little interest in regulating transnational capital for the common good. Instead, they have increased the power and growth of the finance sector and of transnational corporations by moving them out from under the authority of nation states and into the emerging corporate feudalism of an open access global commons.

2.6.3 Charles Darwin Meant Local Competition, Not Global Trade

Adam Smith and David Ricardo are not the only intellectual giants of the European past whose theories have been grievously simplified and misquoted. Charles Darwin, termed one of the most scientifically influential human beings who ever lived, postulated explanations of the origins and progress of life that have formed the basis for all the modern life sciences. His name and theories have been hijacked into the service of economic and social theory, often under the name of 'Social Darwinism'. One of the most disgusting phenomena to emerge was the Nazi ideology postulating a relentless competition for survival among human races.

Darwin's theory surely was built on the observation of competition among species. That competition, however, was mostly a local phenomenon. He knew from the taxonomy of Linnaeus and others that species diversity related to a diversity of locations and habitats. He visited the Galápagos Islands and there found an astonishing diversity of finches evolved seemingly from one pair of finches stranded a few millions of years earlier (Fig. 2.5). This was the final evidence for him to complete his 'Origin of Species'.[35] He clearly saw that it was the *absence* of non-finch competitors on the islands that allowed finches to explore and conquer new niches and thus to evolve into new species.

[35] Darwin (1859).

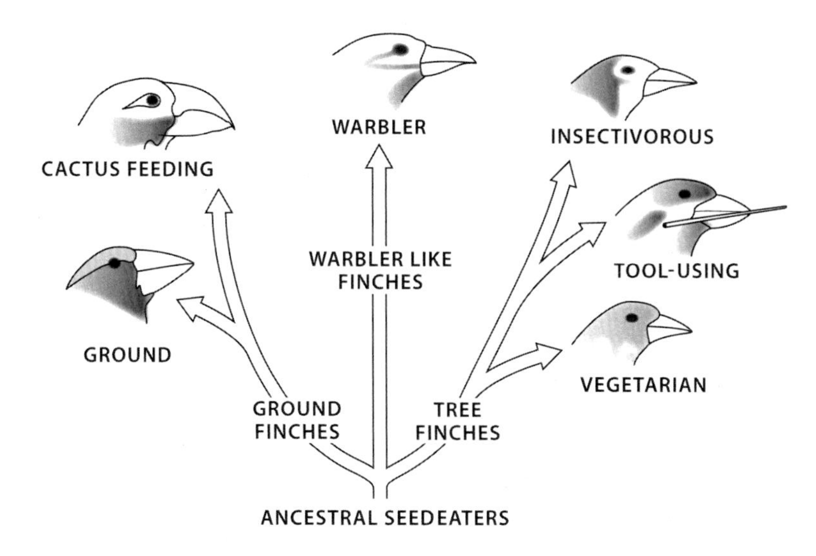

Fig. 2.5 Darwin finches on the Galápagos Islands originating from one ancestor pair evolving into many different specializations (and species) (After: www.yourarticlelibrary.com/evolution/notes-on-darwins-theory-of-natural-selection-of-evolution/12277)

Modern population Darwinism, developed by J.B.S. Haldane, Ronald Fisher, Theodosius Dobzhansky and others,[36] has established another astounding evolutionary feature of *limiting* competition. The basis was a phenomenon, well known since Gregor Mendel's discoveries in the nineteenth century, that genes come in pairs ('alleles'), of which one tends to 'dominate' the other, the 'recessive' allele. Recessive traits of the 'genotype', the genetic outfit of individuals, tend to remain invisible in the 'phenotype' the expression in the physical body of the genetic outfit. The brown iris in humans is dominant over the blue iris. You can't tell by looking into somebody's brown eyes whether she or he carries a blue iris gene from mother or father. However, blue-eyed persons are definitely 'homozygote' (double) carriers of the 'blue iris' gene from both parents.

Differing eye colours are conspicuous traits, called *mutations* at the first instance of their occurrence. Conspicuous mutations were the basis of Gregor Mendel's experiments with peas and other species. But in the real world, they are the exception. The rule is tiny genetic mutations, which are usually recessive and therefore remain 'hidden' under their dominant wild-type alleles. This mechanism, as Haldane et al. realized, allows for the accumulation over millennia of huge 'gene pools' containing very large numbers of mutations. Most of them are not only recessive but would be less fit than the respective 'wild types' if expressed in the phenotype (i.e. if inherited from both parents). However, being recessive, *they remain protected*

[36] E.g. Dobzhansky (1937). Or: Huxley (1942).

against selection for long time intervals because the statistical probability of them being present in both parents is always very small.

Population biologists of the 1930s explained the mechanism as a realistic basis for *continuous* and adaptive evolution. They argued that a small but relevant statistical probability could bring two equal recessive parental genes together, and another probability would make the respective phenotypes the right answer to changing environments. Evolution would no longer depend on the surfacing of 'hopeful monsters', conspicuous mutations, that were the subject of speculation when biologists tried to bring Darwin's theory together with Mendel's findings. The concept of the gene pool gave Darwinism its plausibility back. It explained the positive evolutionary value of protecting and accumulating *less fit* traits, even such things as hereditary diseases, like the genetic propensity for a few populations of humans to develop sickle-cell anaemia; that gene also confers some immunity from local infections like malaria.

Some evolutionary biologists, however, notably agricultural breeders struggled with and hence disliked the abundance of invisible recessive genes finding them an obstacle to strategic breeding. They wanted homogeneity, not diversity of the genes. But then, such homogeneous, domesticated varieties tend to be less robust, less able to adapt to unforeseen challenges of weather, nutritional variance and microbial infections. Later scientists, most prominently Stephen Jay Gould and Niles Eldredge,[37] conjectured another important property of the gene pool: rare recessive genes become visible when the interbreeding population is small. This typically happens under unusual stress from new parasites, droughts, or shortage of nutrition. With a small but relevant probability, a few of the recessive mutations may turn out to be a good answer to new challenge, such as resistance to the parasite, lesser need for water, or the ability to adapt to other nutritional sources. In such cases, the advantage of the recessive gene soon becomes an advantage for the whole population: another proof of the usefulness of protecting options that under earlier conditions would have been inferior.

Perhaps the most up-to-date summary of evolution according to Darwin is by Andreas Wagner.[38] His mantra is the build-up over millions of years of immense 'libraries' of genetic options in each species. Until recently, these libraries were erroneously called 'junk DNA'. In reality, species can use their library for trying out lots of *existing* genes that are coding for hopeful proteins. Building up new genes and proteins from scratch would take a great deal more time. So Wagner convincingly says that innovations of evolution are dependent on their library, a treasure-trove that should be protected against destruction by natural selection, even if most of the 'books' of that library may be 'inferior' compared with the related 'books' of momentary rivals.

At this stage, it seems timely to mention a spectacular new discussion in genetics, usually called 'genome editing' or CRISPR/Cas9-system (*Clustered Regularly*

---

[37] Eldredge and Gould (1972).

[38] Wagner (2015).

Interspaced Short Palindromic Repeats). Developed and published in 2012, the method allows technicians to cut and modify DNA at specific sites, thus potentially cutting out disease-causing genes.[39] The scientific community, especially medicine, is thrilled by such potential. A recent NAS report is full of optimism for applications in public health, ecosystem conservation, agriculture and basic research.[40] Critics, for example, from the innovation watchdog NGO ETC group,[41] say that the partly defence-funded report fails to address the three major concerns of genome editing: militarization, commercialization and food security.

Even in the scientific community, however, hesitation prevails for the time being in terms of applying this method to human genome editing. If this methodology becomes mainstream, one would expect/fear a systematic narrowing down of genetic diversity, reducing the size of Wagner's 'library'. A minimum period of caution and further research must be implemented, so as not to systematically disrupt the inherited diversity of any species being subjected to genome editing.

In any case, we are learning that to properly understand Darwinism, it is essential to recognize that *limiting competition* and the *protection of weaker strains* are both indispensable pillars of evolution.

By contrast, doctrinaire economic theory assumes that innovation and evolution always benefit from high-intensity competition everywhere and from eliminating the weak – a simplification that is virtually the opposite of the truth.

2.6.4 Reduce the Contrasts

Three times this analysis said 'by contrast'. Many of the problems in modern economic theory may lie in its use of false or exaggerated quotations of, or references to, three of the giants of economic and social thought. Darwin of course did not see himself as a father of economics, but his discovery of the power of competition and selection is fundamental to the concept of markets. Adjusting the currently misleading quotations and references could run as follows:

- The blessings of the *invisible hand* require the existence and effectiveness of a strong legal framework that must lie outside the influence of powerful market players.
- The theory of comparative advantage does not automatically apply to trade in *capital*. The power of capital is dangerously asymmetric: Big capital will always have a 'comparative advantage' over small capital; and many local innovations only need small capital.

---

[39] For an early but comprehensive description including of applications, see Hsu et al. (2014).

[40] NAS (2016).

[41] An Action group originally on 'Erosion, Technology and Concentration' (ETC). It works to address the socio-economic and ecological issues surrounding new technologies impacting the world's most vulnerable people.

- Competition in its origin is a localized phenomenon. Protecting – to an extent – local cultures, local specializations and local politics against the immense powers of world-size players can be helpful for diversification, innovation and evolution. The term 'non-discriminatory' was originally a good anti-racist term but is currently[42] used in trade agreements with a view to exclude protection measures for weaker and local producers.

This quick treatment of the some failures of doctrinaire economics can surely be improved. For the present, however, many economists, historians and academics, as noted in our references, are in agreement that this general line of thinking is correct. Such a list could become a powerful and persuasive reason for overhauling the doctrine that growing numbers of people find so worrisome and unjust.

Fortunately, a strong movement has emerged demanding pluralism in economic teaching. It is called ISIPE[43] and originated in Paris but now consists of more than 165 associations in more than 30 countries. They demand, among other things, that 'the real world should be brought back into the classroom, as well as debate and a pluralism of theories and methods'.

Some prominent economists are supporting the movement for systemic economic change and are also closely associated with the Club of Rome. They include Robert Costanza and Herman Daly, Tim Jackson,[44] Peter Victor[45] and Enrico Giovannini[46] (who has contributed on the statistics for a post-GDP definition of well-being).

2.7 Reductionist Philosophy Is Shallow and Inadequate

2.7.1 Reductionist Philosophy

According to the findings of Sects. 2.4 and 2.5, the economic philosophy of the superiority of markets has become the worldwide *dominant* paradigm rather recently, after the end of the Cold War. But that market doctrine has been failing to a relevant extent, and some of the core tenets of today's market philosophy consist of massive misquotes and misunderstandings of their original meaning. That should lead to a better understanding of the *philosophical* errors of the doctrine, and to a broader mind-set. For this, one has to dig deeper into the philosophy of understanding, into the epistemology of human existence and of nature.

Over the centuries, researchers in the natural and social sciences grew to believe that the more detailed their descriptions became and the more they could dissect the

[42] Originally, the concept of non-discriminatory was introduced in the context of human rights, protecting the weak!

[43] ISIPE (International Student Initiative for Pluralism in Economics) 2014; www.isipe.net, Open Letter.

[44] Jackson (2009).

[45] Victor (2008).

[46] Giovannini et al. (2007). See also OECD (2015).

elements into smaller parts, the more progress they would make. Since Descartes and Newton, some kind of *hierarchy of exactitude* developed. Mathematics was at the top of that ladder. John Locke, in a learned letter of 1691, made the statement that 'The price of any commodity rises or falls by the proportion of the number of buyer and sellers'. This was the first mention of what became known as 'the law of supply and demand', an accepted formula for determining the price of goods and services. Locke's statement was very close, deliberately, to Isaac Newton's contemporary third law of motion, stating that every action has an equal and opposite *reaction*. Physics and economics seemed to resemble each other; both would pride themselves as being scientifically exact and beyond normative quarrels.

Authoritarian doctrines based on power, faith or superstition had plagued people (and scientists in particular) for so long that finding methods leading to hard facts must have appeared as a great liberation to them. Logical induction and empiricism brought the notions of evidence and quantitative measurement alongside the idea of objectivity. All of these have been enormously useful; skyscrapers and airplanes, computers and EKG machines are all manifestations of this form of scientific research and development.

However, not all studies are served by the empirical and inductive quantitative method. There are some forms of understanding that resist measurement, that elude objectivity and that are outside of the reach of criteria that is great for building rockets, and not useful at all for raising children or understanding the ramifications of culture on climate change. Reductionism, or the habit of isolating information from its context(s), has been good to us, and it also has been deadly.

The reductionist approach has limits even for describing and understanding biological and human reality. One is that facts, by definition, refer to the past, not the future. Another is that laws of physics and mathematics may be unambiguous and strict, while facts in economics tend to be approximations only. For example, physical force and counterforce (Newton) can be assumed to be equal under all conceivable circumstances, while supply and demand can differ, can change and can be influenced by fashion, moral attitudes or lowering or raising prices. Finally, the act of measurement can interfere with the previous facts and thereby change them, even in physics, such as in the extremely surprising discovery of Werner Heisenberg in 1927,[47] who called it the *uncertainty relation*. Niels Bohr, hearing about this discovery, said that the uncertainty principle was a manifestation of a more profound principle, called *complementarity*: two complementary properties cannot be measured exactly at the same time.

Facts, derived in terms of complexity are not hermetic. The observer matters, and the teams of observers matter. Since data are always derived through the particular lens of the researchers, descriptions of their filters of perception are vital information not to be sterilized out of findings. All this should be commonplace in the social sciences but also in medicine and technology. For engineers and medical doctors, succeeding in *changing* previous facts is actually the intended objective. In physics,

[47] E.g. Heisenberg (1930).

however, the master discipline of exactitude, Heisenberg's principle came as a huge surprise, a sea change for the understanding of measurement.

It also served as a shock for *analytic philosophy*, which had emerged, chiefly in the Anglo-Saxon world, as the ostensibly adequate, perhaps the *only* adequate philosophy of science. It is also characterized by ever more precise measurements of ever smaller entities: The great advances of physics were closely associated with the understanding of atoms and elementary particles. The discovery in 2012 of the Higgs boson was celebrated as some kind of a coronation of physics. Likewise, modern biology to a very large extent became *molecular* biology. Agricultural sciences proceeded from plant and animal biology to the chemistry of soils, fibres and meat, and to chemicals for pest control and nutrients. And economics became ever more mathematical.

As early as the 1930s, however, Niels Bohr insisted that the complementarity principle was not restricted to physics, but was even more valid, as indicated above, for biology, the social sciences, medicine and engineering. The researcher's interference with his object was an essential part of the cognitive act.

That cognitive act found its limits if the interference destroyed some essential features of the object. That is the case, most conspicuously, if a scientific investigation of a living organism involves killing it, that is, destroying the very nature of its life in order to learn about it (see Fig. 2.6 and the box on pollinators).

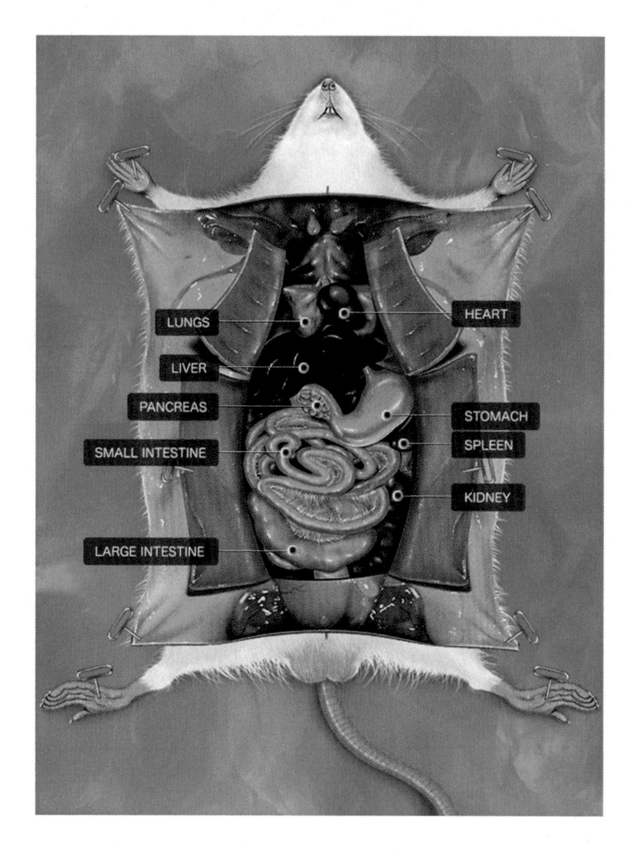

Fig. 2.6 Dissecting a rat means killing it; ironically, this is called life sciences (Source: Emantras Inc., see: www.graphite.org/app/rat-dissection)

The case of pollinators is currently relevant. It is much in the discussion these days, and there is still a chance of preventing mega-disasters. But there is another type of symbiosis, which is at least as important for robust and healthy ecosystems. It's the cooperation of all plants with the hugely variant world of mycorrhizal fungi. It must be raised to equal prominence with the pollinators when it comes to define and effectively protect soil fertility.[48]

Box: Pollinators as Victims of Reductionism

Industrial agriculture is tempted to ignore the unpaid service bees and other pollinators are providing. For millennia, they have carried pollen from the male parts of flowers to the female parts and making them fecund and productive.

In agro-ecosystems, pollinators are essential for orchard, horticultural and forage production, as well as the production of seed for many root and fibre crops. About two-thirds of the crop plants that feed the world rely on pollination by insects or other animals to produce healthy fruits and seeds. Of the many crop species providing basic food supplies for around 150 countries, nearly three quarters are bee-pollinated (not all by honeybees), and several others by thrips, wasps, flies, beetles, moths and other insects.[49]

For human nutrition the benefits of pollination include not just abundance of fruits, nuts and seeds, but also their variety and quality; the contribution of animal-pollinated foodstuffs to human nutritional diversity, vitamin sufficiency and food quality is extremely important.

Current agricultural practices are causing grave threats to pollinators. As farm fields have become larger and less diverse and the use of agricultural chemicals that impact beneficial insects such as pollinators along with plant pests has increased, pollination services are showing alarming declining trends around the world. Chemical pesticides are applied to crops to kill insects that reduce yields, not to impact pollinators. But the toxic chemicals regularly also affect beneficial insects such as pollinators and natural enemies, and in fact over time more often make pest problems worse.

Recent evidence shows that modern insecticides, particularly the neonicotinoids, have lethal and sub-lethal effects on bees, sufficient to impact the ability of the insects to effectively pollinate.[50] This group of insecticides is used now ubiquitously as seed coatings, thus without any of the known good practices of observing first before treating problems. The group of toxicants are *systemic*, which means they are taken up into the young plant and transmitted to all tissues, including pollen and nectar, thus ubiquitous within the field as well as across agricultural landscapes. We have thus evolved our productive landscapes into ones that are killing fields for the ecosystem services that can sustain production.

[48] Quote needed!

[49] Singh et al. (2013). Also: Klein et al. (2007).

[50] IPBES (2016).

This observation interference dilemma makes it plausible that, for the understanding of living systems and other open systems, analytic philosophy might be inadequate. After the early pioneering work of Gregory Bateson,[51] two eminent scientists and philosophers, Fritjof Capra and Pier Luigi Luisi, have undertaken a profound discussion aiming at a new *philosophy of life*, calling it *The Systems View of Life*[52] in their book by that name. It surely includes a harsh critique of reductionist and purely analytical thinking.

In this book, they walk the reader through Newtonian physics and the ensuing mechanistic view of life that led to our current mechanistic social beliefs. The latter they associate closely with the Enlightenment, which was also the period when classical political economy and economic modelling originated. The 'machine metaphor in management' was a major paradigm of the Industrial Revolution and led, in the early twentieth century, to Frederick Taylor's *Principles of Scientific Management*, today termed 'Taylorism'.

After having exposed how the reductionist approach may work for machines but is inadequate for the study of living systems, Capra and Luisi proceed to early systems theory, such as the cybernetics of the 1940s, the autopoiesis thoughts of Maturana and Varela,[53] and the principles of nonlinear (i.e. jump-prone) dynamics. They offer an excellent interpretation of Charles Darwin's 'tree of life' (published 22 years before his *Origin of Species*) to demonstrate that 'we are not our genes',[54] a position that squarely contradicts populist ideas concerning genetics and race.

In a scientifically transparent and persuasive pursuit of their 'systems view of life', the authors say that science and spirituality should not conduct a 'dialogue of the deaf',[55] staying in their separate silos, but should seek and develop their mutual parallels. Echoing Benedictine monk David Steindl-Rast,[56] they praise 'spirituality as common sense'; but in line with Fritjof Capra's earlier work, they observe that Asian religions and spirituality have a much more direct way of finding synergies with modern science than do the monotheistic religions, notably Islam, Judaism and Christianity.

A similar approach to understanding living systems and to overcoming destructive arrogance comes from biologist Andreas Weber in his new book *The Biology of Wonder*.[57] The disconnect between humans and nature, he says, is perhaps the most fundamental problem faced by our species today. This book demonstrates that there is no separation between human beings and the inhabited world, and in so doing, it validates the essence of our deepest experiences, especially in nature. By reconciling science with meaning, expression and emotion, this unusual work helps us to

---

[51] Bateson (1979).

[52] Capra and Luisi (2014).

[53] Maturana and Varela (1972).

[54] Capra and Luisi, l.c., p 204–207.

[55] Ibid. p. 282–285.

[56] Steindl-Rast (1990).

[57] Weber (2016).

better understand where humans stand within the framework of all other life. Contrary to the dissections and experiments of modern biology, this approach emphasizes the mutually dependent relationship between the life sciences – and life.

Readers having followed the considerations so far may be willing to agree that a reductionist philosophy is not only inadequate for dealing with living systems but also for overcoming the tragedies of a destructive socio-economic development in the 'full world'.

2.7.2 *The Misuse of Technology*

Another example of the limitations of reductionism – and of the way science is organized – is our current inability to assess the wider implications of the technological revolution. We are living in an age when science and technology are undergoing dramatic changes, notably in areas like artificial intelligence (AI), robotization and nanotechnology, as well as in the bio- and neurosciences. This means that predictions of the technological landscape only 5–10 years ahead are very difficult to make.

Many new opportunities will open up for products and services that are beneficial to mankind, but the same goes for opportunities for misuse. Let us focus attention on one technological area that is particularly intriguing – AI and robotization.

In his most recent books, *Sapiens* and *Homo Deus*, Yuval Harari[58] pays particular attention to robotization and AI, as both include some kind of marriage between man and machine. Harari paints a dystopic vision of a future in which humanity is increasingly dominated by intelligent machines. He even suggests that with the rise of brain-machine interfaces, today's elite classes might upgrade themselves to a biologically improved version – super humans – which the general masses couldn't possibly afford. The result would be a caste system with real biological hierarchies.

We are at a critical fork in the road, and Harari asks us to stop and reflect. The very essence of who we are as humans is at stake. Like him, our contention here is that most people – including policymakers – are unaware of the dangers posed by AI. Neither at the international nor at the national levels do we have institutions whose task is to broadly follow scientific and technological developments and assess their implications, and in particular their risks. The digital economy is often discussed, but primarily in terms of how it will affect productivity and competitiveness.

When it comes to robotization and AI, most of us are ready to accept that computer programs can beat human chess masters; computers are far more efficient calculators than humans. At the same time, most people believe that there are limits for IT. Computers, it is argued, can never rival humans in the arts, because art requires something distinctively, perhaps even spiritually human, which can never be replicated by computers. However, Harari brings to his readers' attention

[58] Harari (2011, 2017).

a computer program written by a professor of musicology that produced a musical composition that an audience thought was superior to Bach.

If technology can perhaps outperform people in our allegedly distinct 'human art' forms, there's really no reason to think that it can't outperform humans in every other field. Computer programs may lack subjective consciousness like ours – and that is an important distinction. However, that doesn't prevent them from outperforming us in both the intellectual and artistic fields.

What kind of 'algorithms' does Harari have in mind? These of course currently happen to be written by humans. The first kind, encoded within computational machines, will create new technological beings with artificial intelligence. The second, encoded in DNA, will create new biological beings with higher 'natural' intelligence. The growing technological capacity to manipulate two fundamental forms of information – the biological and the computational, the byte and the gene – will thereby result in the birth of superior beings; and it is hard to imagine such beings will not ultimately overrun our world. They will take over our jobs, infiltrate every aspect of our lives, even control our emotions and influence our opportunities, as easily as they control our traffic and bank machines today.

In conclusion, the issues Harari raises are of profound importance. All of human society needs to participate in a thorough discussion about where information technology might take us in terms of its opportunities and its threats. The implications in terms of disruptive change are significant in many areas – global effects on employment, privacy and, indeed, who can be defined as 'human' in the future. The possible marriage between man and machine opens up questions that are hugely challenging. To address them, policymakers need to anchor science and technology more firmly within ethical boundaries. We must also develop institutions capable of continuously assessing the broader implications of such sweeping technological developments.

2.8 Gaps Between Theory, Education and Social Reality

A reductionist philosophy as described in the last section and our growing tendency to divide reality in order to apprehend it led to a divorce between our knowledge, our educational system and the social reality that we live in. Not only has the ivory tower of academia been far away from social reality for many centuries but also has the increasing reductionist tendency led to ever more specialized disciplines. Human endeavour has moved away from looking at reality as a whole to cutting it up in small pieces – a major source of the challenges we are facing.

The division between economy and ecology that persisted for nearly two centuries is a dramatic example of a general problem. Fragmentation of knowledge results in a loss of perspective regarding the interrelationships and interdependencies between the parts and the wider whole of which they are components. This disconnect has shaped the organization of universities and research institutes on narrow disciplinary lines. It has shaped the organization of government of policy and administrative functions in increasingly specialized areas of expertise divorced from the wider social reality of which they are parts. It has resulted in laws and policies that seek to address specialized issues, unmindful of their impact on other fields. It is the reason why until the 1970s, environmental impact studies were never considered important in planning for new commercial or public projects.

This tendency also contributes to the growing divorce between financial markets and the real economy, between technology and employment and between economic theory and public policy. As a consequence, financial markets have acquired an independent life of their own, mathematically rather separate from the real economy they were created to support. The trend of technology to increase human productivity has now become development of technology for its own sake, steered by cost-effectiveness – at the expense of employment and welfare.

The impact of this divorce on policy and practice has been equally apparent and influential at the theoretical level. It led to a fragmentation of economic theory into countless sub-disciplines and the formulation of countless theories and models confined to describing the internal operations of a specific type of activity, while marginalizing important factors as mere externalities. The same tendency facilitated economics in developing separately and independently of the other social sciences. The physicists who founded neoclassical economics in the late nineteenth century also played a part, determined as they were to elevate economics to the status of a true science based on physics.[59] These trends resulted in the prevalence of economic theories that show mathematical virtuosity while partly ignoring political, legal, social, cultural and psychological factors of relevance to an understanding of human economic behaviour and systems.

The same fragmentation of disciplines has occurred in all the natural and social sciences, with similar consequences. Specialization triumphs over integral views.

Models and theories play a vital role in education, but when the theory is not relevant to social reality, or the model is mistaken for reality, the resultant education creates a workforce and citizenry that are grounded in book learning, but not suited for the real world. Tomas Björkman, an economist and former investment banker, traces the consequences of this divisive tendency on economic theory and models. He has identified three gaps between theory and practice. The first is the gap between the myth (our common understanding) and the model: In Björkman's words, 'there is a huge gap between what the *model* is actually telling the economists and what we as the general public tend to believe it is telling us. Such beliefs are often not explicit, as we have seen, but creep in as underlying assumptions in language and policy making'. The second gap is between the actual real-world market and the neoclassical model. Economists know that the neoclassical model is not based on the real market but on a set of very limiting assumptions of a 'perfect theoretical market', depicting the market in a perfect state of equilibrium. In this way, reality can be expressed in equations that have little to do with humans, their institutions, or their innate potentials, aspirations, emotions, values etc. The third gap is between the prevailing market and the possible market – market as it could be. This myth has its roots in the assumption that the market is a fixed reality.[60]

[59] Léon Walras and William Stanley Jevons.

[60] Björkman (2016).

Recent research shows how unrealistic these assumptions are, but economists and economic students continue to study these models and mistake them for an adequate representation of reality. An economic model in general is not designed to model the actual world. It is designed as a way to investigate what insights theoretical assumptions and abstractions might lead to. A model may assume the availability of perfect competition, market information and predictive capacities. But in the real world, perfection is an ever-receding goal. Such a gap between academic learning and the needs of the real world translates into wider gaps in all spheres. For further elaboration, see Sect. 3.19.

2.9 Tolerance and Long-Term Perspectives

The philosophical crisis has massive effects on the way the world is governed. In this section, only some of the major features of a 'philosophy' for a sustainable world will be discussed.

One important aspect has so far been completely neglected. It is the relation between national and global governance. The United Nations was created at a time when the nation state was the only entity entitled to adopt legally binding laws. Sure, there were also legally binding international treaties, but they were meaningless unless ratified by all nation states involved.

Under the impressions, terrible as they were, of the Second World War, fought by autonomous nation states, if connected in various military alliances, the idea became politically arguable that the age may have come where nation states are losing some of their earlier sovereign rights, including complete military sovereignty.

When the world takes the sustainability challenges serious, many other areas of sovereignty have to be put in question. This requires the establishment of a truly new mind-set. The climate discussions are the most familiar case in point. It is becoming increasingly *immoral* to continue blasting greenhouse gases into the atmosphere. And yet national parliaments and governments, not to speak of nationalistic citizens' movements, often consider it unacceptable to be forced into national legislation by international institutions or conferences.

The European Union, on the other hand, is a shining example for the fact that surrendering national sovereign rights to a larger authority can actually be beneficial for the countries doing so. In fields like commerce, consumer protection, agriculture and environment, some 80% of the legislation of an EU member state are prescribed by European directives or even directly binding regulations. Of course, there are procedures in which national governments negotiate about the contents of directives and regulations. But increasingly, qualified majorities of countries can outvote minorities of countries still making the EU law binding for all; there are exceptions where unanimity rules, such as issues related to taxation. All economic analyses have shown that the EU countries greatly benefitted from this reduction of their national sovereignty.

On the global scale, nothing of that kind is visible so far. Of course, there is the Law of the Sea, the UN conventions on climate and biodiversity, and many other international legal instruments. But enforcement is hardly possible – with one major exception – the rulings of the WTO (see Sect. 1.9).

It is time to take a new initiative for a more stringent international legal system, adequate for the 'full world'. The very idea of national sovereignty was a product of the 'empty world'. For a further discussion, see Sect. 3.15.

Some religions and cultural traditions stemming from the empty world had a basically intolerant credo legitimizing aggression, expansion and victimization of the people of other creed, colour, or culture. Also this must definitely be overcome.[61]

2.10 We May Need a New Enlightenment

2.10.1 New Enlightenment, Not Renewed Rationalism

In earlier sections the power of the European Enlightenment of the eighteenth century was mentioned. Its strongest figures were David Hume, Jean-Jacques Rousseau, Voltaire, Adam Smith and Immanuel Kant, but it was built on powerful precursor philosophers such as René Descartes, Blaise Pascal, Francis Bacon, Erasmus of Rotterdam, John Locke, Baruch de Spinoza, Montesquieu, G.W. Leibniz and Isaac Newton, to name but a few. Together, they caused and created a revolutionary change in European civilization.

One of the most revolutionary features was the separation of the nation state from the church(es). While the then existing church had little sympathies for independent intellectualism, the enlightened state saw free thinking and acting of citizens as a great hope for the future. They also saw an enlightened citizenry as the prime source of scientific endeavour, technological inventiveness and the entrepreneurial spirit. And in fact, the eighteenth century witnessed an explosive development of science and technology. Antoine de Lavoisier and James Watt were among the first, but after them an avalanche of technological innovations led to the Industrial Revolution.

The Enlightenment is also credited for freeing human *individuals* from the suffocating pressures of churches and the absolutist state of the seventeenth and eighteenth centuries. But this new individualism also led to the gradual decay of earlier communities. The *commons* (like grazing land, forests or fishing grounds) were the foundation of earlier human survival. However, in parallel with growing private wealth and with the new appreciation of individual achievements, these got eroded, privatized and in some cases destroyed.

For civilizations outside Europe, the downsides of the Enlightenment had even worse effects. European armies, colonists and missionaries had already conquered

[61] Koopmans (2015).

and colonized much of the world during the sixteenth and seventeenth centuries, and the Industrial Revolution that followed made Europe, notably the British Empire, essentially invincible. European supremacism and missionary war went as far as justifying the subjection and killing of peoples living in conquered territories. Many alternative traditions and cultures were destroyed, cultures that had existed and developed for thousands of years. Peter Sloterdijk goes so far as to place all the burden for the horrors of European missionary colonialism on the monotheistic religions and parallels this period with the mentality of current Islamic 'holy wars'.[62]

Certainly the European development of rationalism, science and technology was also a vehicle for progress for humanity at large. But what was said in relation to Pope Francis' encyclical *Laudato Sí*, about our current philosophical crisis and the suicidal features of modern capitalism, should lead in the full world to *the demand for a new kind of Enlightenment.*

It has actually become fashionable to call for new Enlightenments, but motives and contents differ widely. In many cases, the word is used simply to revive or modernize the old Enlightenment concepts of rationalism, freedom, anti-normative, anti-regulation, anti-state dominance. One example out of many comes from the British Libertarian Alliance.[63] Another has been the 'March for Science' in April 2017, attended by more than a million protesters against President Trump's blatant disrespect for facts. The March emphasized that science upholds the *common good* and called for *evidence-based policy* in the public's best interest.

The reasoning in this Chap. 2 of this book takes a different approach. Sure, rationalism is essential in unmasking 'fake news' and other nasty trends, but it's also obvious that it should not be used to overwhelm cherished and sustainable traditions and systemic values that are not susceptible to anatomic analysis.

The new Enlightenment, 'Enlightenment 2.0', is unlikely to be Europe-centred. It should look at the great traditions of other civilizations. To give two very different examples:

- The Hopi tradition in North America remained essentially stable and sustainable for 3000 years. They are one of the oldest living cultures in documented history. They developed a sustainable kind of agriculture, maintained an essentially stable population size, avoided wars and became masters in building stone houses. Under almost every definition of sustainability, they would be among the champions. Their very complex religion is based on the idea of balance – between resources like water and light, between differing peoples, between day and night and the various seasons and even between humour and sobriety.[64]
- In most of the Asian traditions, there is a strong sense of balance, as opposed to the monotheistic dogmatic view where only one side is right. Balance is sought between rational thinking (the brain) and emotional feeling (the heart).

[62] Sloterdijk (2009).

[63] Tame (1998).

[64] http://hopi.org/wp-content/uploads/2009/12/ABOUT-THE-HOPI-2.pdf

Fig. 2.7 The Yin and Yang symbol

2.10.2 *Yin and Yang*

Yin and Yang are symbols of *balanced contrast*. Mark Cartwright[65] in a contribution to the Ancient History Encyclopedia offers a simplified definition of what is also an essential part of Confucian cosmology:

> The principle of *Yin and Yang* is a fundamental concept in *Chinese philosophy* and culture in general dating from the third century BCE or even earlier. This principle is that all things exist as inseparable and contradictory opposites, for example female-male, dark-light and old-young. The two opposites attract and complement each other and, as their symbol illustrates, each side has at its core an element of the other (represented by the small dots). Neither pole is superior to the other and, as an increase in one brings a corresponding decrease in the other, a correct balance between the two poles must be reached in order to achieve harmony.

Figure 2.7 shows the familiar picture of Yin and Yang.

> Yin is feminine, black, dark, north, water (transformation), passive, moon (weakness), earth, cold, old, even numbers, valleys, poor, soft, and provides spirit to all things. Yin reaches its height of influence with the winter solstice. Yin may also be represented by the tiger, the colour orange and a broken line in the trigrams of the *I Ching* (or Book of Changes).
>
> Yang is masculine, white, light, south, fire (creativity), active, sun (strength), heaven, warm, young, odd numbers, mountains, rich, hard, and provides form to all things. Yang reaches its height of influence with the summer solstice. Yang may also be represented by the dragon, the colour blue and a solid line trigram.
>
> As expressed in the *I Ching*, the ever-changing relationship between the two poles is responsible for the constant flux of the universe and life in general. When there is too great an imbalance between yin and yang, catastrophes can occur such as floods, droughts and plagues.

This brief description cannot, of course, explain the whole richness of the Yin and Yang philosophy, which can also be criticized for typified and hence unfair gender roles, or having the static features of zero-sum games (positive sum games

[65] Mark Cartwright (2012) Yin and Yang. Definition. Ancient History Encyclopedia.

are much preferable). But it reflects a wisdom in the comprehension that contrasts can be creative. This wisdom differs from prevailing Western and Islamic habits that see contrasts as an invitation to decide which one is right (or benign) and which one is wrong (or evil), often leading to bitter and violent feuds. Of course, also Western traditions have embraced balance. Notably the dialectic philosophy of G.W.F. Hegel should be mentioned but cannot be expanded here.

2.10.3 Philosophy of Balance, Not of Exclusion

The wisdom of the synergies to be found between contrasts can also help overcome deficits of the analytical philosophy of science – making room for a more future-oriented philosophy. To be sure, technical and scientific measurements must be done correctly; facts must be treated as facts. But modern physics has shown that the precise measurement of one feature can destroy the measurability of its contrasting (complementary) feature – for example, Heisenberg's *uncertainty relation*, which established that the momentum and position of a particle cannot be measured simultaneously with unlimited precision. The physical basis of this astounding finding lies in the fact that the particle also has wave properties that interfere with the waves (e.g. light waves) of the measuring instrument. Also particle properties and wave properties are mutually 'complementary'.

Complementarity can be a door-opener to perceiving parallels between modern physics and Eastern wisdom and religions. In his bestselling *The Tao of Physics*,[66] Fritjof Capra, mentioned earlier, who was once an academic assistant of Heisenberg, showed that Buddhism, Hinduism and Taoism had the power of dealing with unexplainable realities that people call mysticism. At the end of this book, Capra stated that 'science does not need mysticism and mysticism does not need science, but man needs both'.

Complementarity, balance and the wisdom of synergies between contrasts should be milestones on the way to a new Enlightenment. Surely there will be more philosophical steps for overcoming the deficits of analytical philosophy, of selfishness, individualism, short-termism and other traits mentioned by Pope Francis in *Laudato Sí*, as destructive and suicidal in terms of our Common Home. But there is surely a short list of topics where a *renewed appreciation of balance* is necessary. Most of the topics listed are not new at all, but all of them suffer from a *lack of balance* in our time.

The new Enlightenment should work on a balance:

- *Between humans and nature*: This is one of the core messages of this book. In the 'empty world' it was a given. In the 'full world' it's a huge challenge. Using remaining natural landscapes, waterbodies and minerals chiefly as resources for an ever-growing human population and the fulfilment of ever-growing consumption is not balance but destruction.

[66] Capra (1975).

- *Between short term and long term*: Humans appreciate quick gratifications such as something to drink when thirsty, and quarterly financial reports by publicly traded companies. But there is a need for a counterbalance to ensure long-term action such as policies to re-stabilize the earth's climate. In addition to long-term ethics, short-term incentives will be needed rewarding long-term action.
- *Between speed and stability*: Technological and cultural progress benefits from competition for temporal priority. That is the most important criterion for scientific careers and is essential for commercial success. 'Disruptive' innovations carry tremendous appreciation. But speed by itself can be a horror for the slow ones, for most of the elderly, for babies and for village communities (think of the Hopis!). What is worse is that the current civilizational addiction to speed is destructive to structures, habits and cultures that have emerged under the sustainability criterion. Sustainability, after all, basically means stability.
- *Between private and public*: The discovery of the human values of individualism, private property and protection against state intrusion has been among the most valuable achievements of the European Enlightenment. But in our times *public goods* are much more endangered than private goods. There are dangers to the commons, the public infrastructures, the system of justice and reliable order. Under international competition for lowest taxes (attracting investors) public goods tend to be neglected and under-financed. The state (public) should set the rules for the market (private), not the other way round. Paul de Grauwe and Anna Asbury[67] have lucidly described how history has produced pendulum swings between private dominance and state dominance. But history has not brought about anything like a lasting balance between the two.
- *Between women and men*: Many early cultures developed through wars during which women were chiefly entrusted with caring for the family and men for defence (or aggression). This model is outdated. Riane Eisler in *The Chalice and the Blade*[68] has offered archaeological insights into cultures thriving in partnership models and has also called for *The Real Wealth of Nations*[69] claiming that the conventional (male dominated) 'wealth of nations' is almost a caricature of real well-being. Balance may not be well achieved by pushing and pulling as many women as possible into positions that have been typified as 'male'. Balance may be better attained by changing the typology of the job functions.
- *Between equity and awards for achievements*: Without awards for achievement, societies can get sleepy and lose out in the competition with other societies. But there must be a publicly guaranteed system of justice and equity. Inequity, according to Wilkinson and Pickett, tends to be correlated with undesirable social parameters (see Fig. 2.8), poor education, high criminality, infant mortality, etc.[70]
- *Between state and religion*: It was a great achievement by the European Enlightenment to separate public from religious leadership, fully respecting religious values and communities. This has to be maintained in a balanced man-

[67] de Grauwe and Asbury (2017).

[68] 1987. San Francisco: Harper Collins.

[69] Eisler (2007).

[70] Wilkinson and Pickett (2009).

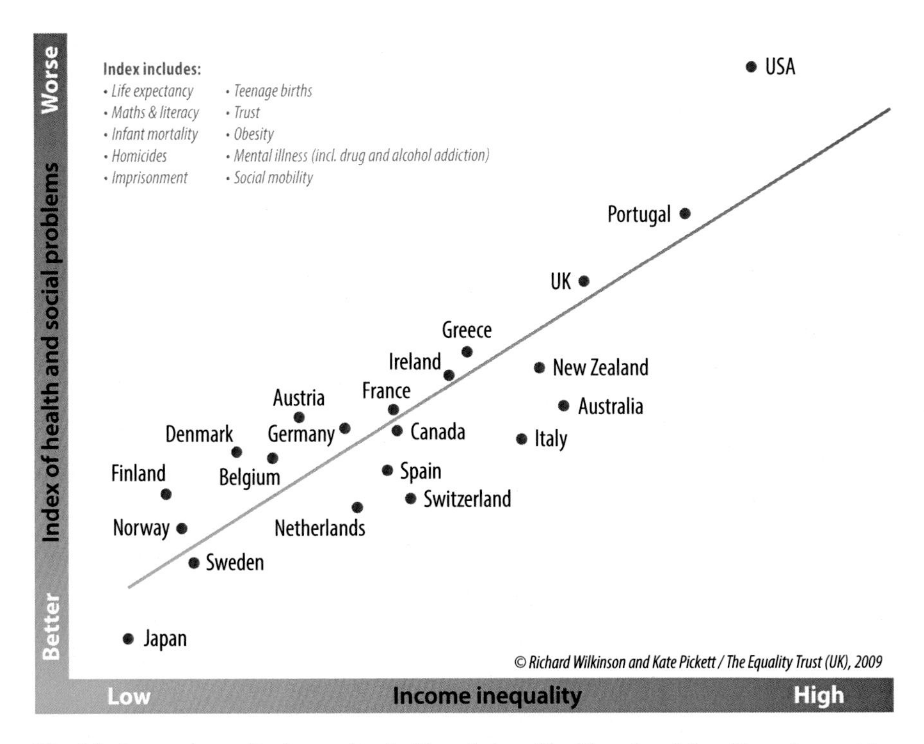

Fig. 2.8 Income inequality is correlated with an index of health and social problems in countries of similar wealth levels (Source: Wilkinson and Pickett [see footnote])

ner. Religions dominating the public sector are in high danger of destroying human rights and the great civilizational achievement of an independent legal system with independent high courts. Dominance of religion tends to be intolerant to persons working outside the religious community. On the other hand, states that are intolerant against religious communities tend to lose touch with ethical (and long-term) needs.

This is a modest and only indicative list of the principle of balance. Lots of other examples of balance would be worth naming and sketching out. They include the dialectic philosophy of G.W. Hegel interpreting human historical developments as thesis, antithesis and synthesis. Ken Wilber (1996), on the other hand, describes the permanent tension between the right and the left halves of the human brain, calling the achievements of the two halves the Two Hands of God. We should repeat, however that balance is just one feature of the New Enlightenment. The list above represents only a modest beginning towards the understanding of balance.

Linking Chapters 2 and 3

Chapters 1 and 2 were descriptive, historical and mostly analytic except, of course, for Sect. 2.10, which contains the proposal of creating a new Enlightenment.

But the world cannot patiently and comfortably wait until all people on earth, or nearly, have adopted a new worldview. That process could take a hundred years, as did the European Enlightenment, which began in the sixteenth century and continued into the era of the French Revolution. New action is required to address the deep troubles of a 'World in disarray' (Sect. 1.1). That's the focus of Chap. 3.

There is actually reason for optimism, because in fact, many trends are moving in the right direction. The German weekly magazine *Der Spiegel* features a series of illustrated articles under the title 'In the old days everything was worse',[71] subverting the familiar complaint that in the old days everything was better. Readers of this series are often surprised to learn that not only weather forecasts have improved and smoking and child labour have receded, but bank robberies, war fatalities, violence,[72] famine, common diseases and religious suppression have also declined. Significantly, however, the entire series' listing of good news is anthropocentric. Examples of ecological good news are restricted to local successes concerning pollution control or species protection, but offer no relief for looming global disasters.

Chapter 3 of this book does present a number of exciting examples showing that local or national actions are possible. They can be and are initiated by states, small communities or businesses at any time. Even individuals can be catalysts for social change, innovation, development and evolution,[73] and this fact is a source of hope. Section 1.1 notes that local experiences often cannot be copied elsewhere. Rules and laws have to be adopted nationally and internationally to slow down and reverse destructive trends and to support sustainable development. But here we are on uncharted ground. This book tries to offer some policy proposals at national levels of course, but also takes the risk of discussing options for global governance, which of course is unavoidable for challenges like global warming.

We therefore invited authors, who are mostly members of the Club of Rome, to present their ideas including on the financial sector, investments and business. That process begins with an essay on a 'regenerative economy', contrasting it with society's current habits of consumption and destruction. We briefly report how two very different countries, China and Bhutan, address the challenges of 'greening' their development. We are interested in legal frameworks that help businesses, including start-up firms, to make money with services and goods that satisfy customers without causing any kind of environmental deterioration. The collection of such thoughts is unavoidably highly diverse and should not be seen as a new doctrine. In the final section, we explicitly invite readers and other discussants to engage and help overcome doctrines and the associated widespread feeling of despair. The meaning of the book's title *Come On!* switches to the encouragement to 'join us on an exciting journey towards a sustainable world'.

Optimism can be seen as the common denominator that includes Chap. 3. This may be at variance with the somewhat 'pessimistic' image the Club of Rome acquired after the publication of *The Limits to Growth*; but another common denom-

[71] Summarized in a book by Mingels (2017).

[72] On the decline of violence, see also Pinker (2012).

[73] Natarajan (2014).

inator might be 'managing for the seven generations to come', which is a principle enshrined in the constitution of the Iroquois Confederacy of eastern North America. Admittedly, that principle may have been easier to honour in the empty world than it is in the full world.

References

Armstron K (2014) Fields of blood: religion and the history of violence. Knopf, Toronto

Baldwin R (2010) Thinking about offshoring and trade: an integrating framework. http://voxeu.org/article/thinking-clearly-about-offshoring

Bateson G (1979) Mind and nature, a necessary Unity. Bantam Books, New York

Björkman T (2016) The market myth. Cadmus 2(6):43–59

Brynjolfsson E, McAfee A (2014) The second machine age. Work, progress, and prosperity in a time of brilliant technologies. W.W. Norton, New York

Capra F (1975) The Tao of physics. In: An exploration of the parallels between modern physics and eastern mysticism. Shambhala, Boulder

Capra F, Luisi PL (2014) The systems view of life. Cambridge University Press, Cambridge

Daly H, Farley J (2004) Ecological economics (chapter 18). Island Press, Washington, DC

Darwin C (1859) The origin of species by means of natural selection. John Murray, London

de Grauwe P, Asbury A (2017) The limits of the market. The pendulum between government and market. Oxford University Press, Oxford

Dobzhansky T (1937) Genetics and the origin of species. Columbia University Press, New York. (3rd revised ed. 1951)

Eisler R (2007) The real wealth of nations: creating a caring economics. Berrett-Koehler, San Francisco

Eldredge N, Gould SJ (1972) Punctuated equilibria: an alternative to phyletic gradualism. In: Schopf TJM (ed) Models in paleobiology. Freeman Cooper, San Francisco, pp 82–115

Francis P (2015) Encyclical Letter Laudato Si' of the Holy Father Francis 2015 On Care For Our Common Home (official English-Language text of encyclical), Vatican

Giovannini E, Hall J, d'Ercole MM (2007) Measuring well-being and societal progress. OECD, Paris

Harari YN (2011) Sapiens. A brief history of humankind. Harper Collins, New York

Harari YN (2017) Homo deus. A brief history of tomorrow. Harper Collins, New York

Harris R (1997) The plan to end planning: The founding of the Mont Pèlerin society. National Review, June 16

Heisenberg W (1930) The physical principles of the quantum theory. University of Chicago Press, Chicago

Higgs K (2014) Collision course: endless growth on a finite planet. MIT Press, Cambridge, MA

Hsu PD, Lander ES, Zhang F (2014) Development and applications of CRISPR-Cas9 for genome engineering. Cell 157(6):1262–1278. ISSN:1097-4172

Huxley J (1942) Evolution: the modern synthesis. Allen & Unwin, London

IPBES (Intergovernmental Science-Policy Platform on Biodiversity and Ecosystem Services) (2016) Summary for policy makers: pollination assessment. IPBES Secretariat, Bonn

Jackson T (2009) Prosperity without growth: economics for a finite planet. Earthscan, London

King A, Schneider B (1991) The first global revolution. A report by the Council of the Club of Rome. Pantheon, New York

Klein A-M, Vaissiere BE, Cane JH, Steffan-Dewenter I, Cunningham SA, Kremen C, Tscharntke T (2007) Importance of pollinators in changing landscapes for world crops. Proc R Soc B Biol Sci 274(1608):303–304

Koopmans R (2015) Religious fundamentalism and hostility against out-groups: a comparison of Muslims and Christians in Western Europe. J Ethn Migr Stud 41(1):33–57

Korten D (2014) Change the story, change the future. A report to the Club of Rome. Berrett-Koehler, Oakland
Maturana H, Varela F (1972) Autopoiesis: the organization of the living. In: Maturana and Varela (1980) Autopoiesis and cognition. Reidel, Dordrecht
Maxton G, Randers J (2016) Reinventing prosperity. Managing economic growth to reduce unemployment, inequality, and climate change. Greystone Books, Vancouver/Berkeley
Mingels G (2017) Früher war alles schlechter. DVA, Munich
NAS (The National Academy of Sciences, Engineering, Medicine) (2016) Gene drives on the horizon. Advancing science, navigating uncertainty, and aligning research with public values. National Academies Press, Washington, DC
Natarajan A (2014) The conscious individual. Cadmus 2(3):50–54
Obermayer B, Obermaier F (2016) The Panama papers: breaking the story of how the rich and powerful hide their money. Oneworld, London
OECD (2015) Measuring well-being. OECD, Paris
Ostry JD, Loungani P, Furceri D (2016). Powerful Hide Their Money IMF's periodical. Finance and Development, June, pp 38–41
Philippe BUC (2015) Holy war, martyrdom, and terror, Christianity, violence, and the west. Univ of Pennsylvania Press, Philadelphia
Piketty T (2014) Capital in the twenty-first century. Harvard University Press, Cambridge/London. (French original: 2013)
Pinker S (2012) Better angels of our nature. Penguin, London
Raskin P (2014) A great transition? Where we stand. Great transition initiative. Tellus Institute, Boston
Raskin P, Banuri T, Gallopín G, Hammond A, Swart R, Kates R, Gutman P (2002) Great transition: the promise and lure of the times ahead. Tellus, Boston
Ricardo D (1951) In: Sraffa P (ed) Principles of political economy and taxation, vol 11. Cambridge University Press, Cambridge, p 136
Singh RP, Prasad PVV, Reddy KR (2013) Impacts of changing climate and climate variability on seed production and seed industry. In: Sparks DL (ed) Advances in agronomy, vol 118. Academic, Oxford, p 79
Sinn HW (2003) The new systems competition. Wiley-Blackwell, Oxford, p 2003
Sloterdijk P (2009) God's zeal: the battle of the three monotheisms. Wiley, Hoboken. ISBN: 978-0-7456-4506-3
Steindl-Rast D (1990) Spirituality as common sense. The Quest 3:2
Stiglitz J (2012) The price of inequality: how Today's divided society endangers our future. W. W. Norton, New York
Tame CR (1998) The new enlightenment: the revival of libertarian ideas, Philosophical notes, vol 48. Libertarian Alliance, London. ISSN: 0267-7091. ISBN: 1-85637-417-3
Victor P (2008) Managing without growth. Slower by design, Advances in ecological economics. Edward Elgar, Cheltenham
Vitali S, Glattfelder J, Battison S (2011) The network of global corporate control. PLoS One 6(10). https://doi.org/10.1371/journal.pone.0025995
von Weizsäcker EU, Young O, Finger M (2005) Limits to privatization. How to avoid too much of a good thing. A report to the Club of Rome. London, Earthscan
Wagner A (2015) Arrival of the fittest. Penguin Random House, New York
Weber A (2016) The biology of wonder. Aliveness, feeling and the metamorphosis of science. New Society Publishers, Gabriola
Weeramantry CG (2009) Tread lightly on the earth: religion, the environment and the human future. Stamford Lake, Pannipitiya
Wijkman A, Rockström J (2012) Bankrupting nature: denying our planetary boundaries. A report to the Club of Rome. Earthscan, London
Wilber K (1996) A brief history of everything. Shambala, Boston, pp 129
Wilkinson R, Pickett K (2009) The spirit level. Why greater equality makes societies stronger. Bloomsbury, New York
Ziegler J (2014) Retournez les fusils!: Choisir son camp. Le Seuil, Paris

Chapter 3
Come On! Join Us on an Exciting Journey Towards a Sustainable World!

3.1 A Regenerative Economy

Humanity is racing with catastrophe. Total system collapse is a real possibility. The evidence of human impact on the planet is undeniable. Radioactive residue of atmospheric testing is now found in geologic deposits. Human releases of CO_2 from fossil fuel combustion have changed atmospheric and oceanic chemistry.[1] So let's not kid ourselves: We face a daunting array of challenges driven by a still rapidly growing population, overuse of resources and the resulting pollution, loss of biodiversity and the declining availability of life support systems.

These stem mostly from the ideological belief that failure to maintain exponential growth in GDP will result in economic collapse. That belief is sewn into the mental models of almost everyone in academia and policy (see the Anthropocene picture 1.6 in Chapter 1.4). But it is wrong. GDP measures nothing more than the speed with which money and stuff pass through the economy (see Sect. 1.12.2).

3.1.1 A New Narrative[2]

It may be grimly satisfying to say there's nothing I can do about it at all, so I'll quit trying and join the party, like the Dark Mountain project suggests.[3] This would be

[1] Vaughn (2016).

[2] The paragraphs above are excerpts from a forthcoming book, "A Finer Future is Possible", by five members of the Club of Rome - Hunter Lovins, John Fullerton, Graeme Maxton, Stewart Wallis and Anders Wijkman - with Hunter Lovins as lead author. One of the objectives of the book is to try to respond to a most challenging question: "Is fundamental change possible without collapse?" The answer from the authors is a resounding Yes and below follows a brief overview of the contents of the book.

[3] Dark Mountain, http://dark-mountain.net/

© Springer Science+Business Media LLC 2018

E.U. von Weizsäcker, A. Wijkman, *Come On!*,
DOI 10.1007/978-1-4939-7419-1_3

the most profoundly irresponsible thing one could do. We are the result of 2 billion years of evolutionary history. We need to act like it.

Further, it is intellectually dishonest. There is a route forward to a finer future. It is, therefore, the obligation of us all to try to create that better world. It is possible for humanity to avoid collapse. But for this to happen, there is one thing more important than anything else: a new narrative to counter the one that placed us on the speeding bus. That was also the basic philosophy of Chap. 2.

The neo-liberal narrative has brought humanity to the verge of ruin. But the Keynesian narrative, if allowed to continue, would drive much the same outcome. It lessens inequality, but would drive the overuse of resources just as badly.

A new narrative could tell us how to achieve a flourishing life within ecological limits; deliver universal well-being, meeting the basic needs of all humans; and deliver sufficient equality to maintain social stability and provide the basis for genuine security.

So what would, in Buckminster Fuller's words, 'a world that works for 100% of humanity' look like? What would it feel like to live there? Our movies are mostly apocalyptic. We know in detail how to battle zombies. We've put a man on the moon. But we have no idea how men and women can walk in happiness on earth.

For the Club of Rome and for many other people as well, the crafting of the basic principles of a new narrative is a first priority.

Dana Meadows taught us: 'People don't need enormous cars; they need respect. They don't need closets full of clothes; they need to feel attractive and they need excitement, variety, and beauty. People need identity, community, challenge, acknowledgement, love, joy. To try to fill these needs with material things is to set up an unquenchable appetite for false solutions to real and never-satisfied problems. The resulting psychological emptiness is one of the major forces behind the desire for material growth. A society that can admit and articulate its nonmaterial needs and find nonmaterial ways to satisfy them would require much lower material and energy throughputs and would provide much higher levels of human fulfilment'.[4]

Our current economic narrative extols competition, perfect markets and unfettered growth in a world in which the rugged individual is seen as the economic epitome. The result is huge inequality. 'Too big to fail' crushes local self-determination and millions of people reportedly hate their jobs. The annual Gallup Healthways survey of US worker satisfaction warns that those are unhappier than at any time measured.[5]

Pope Francis warned that 'The external deserts in the world are growing, because the internal deserts have become so vast'.[6] He also quotes the Earth Charter that challenges humanity, 'As never before in history, common destiny beckons us to seek a new beginning… Let ours be a time remembered for the awakening of a new reverence for life, the firm resolve to achieve sustainability, the quickening of the struggle for justice and peace, and the joyful celebration of life'.[7]

[4] Meadows et al. (1992).

[5] Mann and Harter (2016).

[6] Pope Francis. 2015. l.c., paragraph 217, there quoting his predecessor Pope Benedict XVI.

[7] Ibid., paragraph 207.

The new narrative will emphasize the importance of care, respect for human dignity and the scientific evidence that humans only survived when they are able to organize for the common good.[8]

The good life can be taught. Disciplines like positive psychology and humanistic management[9] have leading business thinkers speaking of flourishing, of Conscious Capitalism, of Natural Capitalism, of Regenerative Capitalism and the need for a Big Pivot. Biologists are exploring the 'wood-wide web', the notion that even natural forests are more about communication and cooperation than cut-throat competition. Policy thinkers speak of Better Life Initiatives, of moving beyond GDP and of happiness indexes.[10] An international consortium, *Leading for Well-being*, is framing a new narrative to capture the following concepts:

True freedom and success depend on creating a world where all prosper and flourish. Institutions serve humanity best when they recognize our individual dignity and enhance our interconnectedness. To thrive, businesses and society must pivot towards a new purpose: shared well-being on a healthy planet.

Good lives do not have to cost the earth. In Sect. 3.14, the Happy Planet Index (HPI) discusses a combination of material sufficiency and expressed satisfaction with one's life. Nature is sustainable not because it set out to be, but because it is regenerative.

3.1.2 Natural Capitalism: Arc of Transition

Creating a sustainable civilization will require sensible policy measures. They must be carried out in communities, by NGOs and by engaged governments, especially cities. Also, it cannot succeed without the participation of the business sector. In many cases, rules will have to be defined by the state or the international community.

Fortunately, there is a strong business case for companies to prosper if they engage in cutting their wastes through efficiency, redesign how they make and deliver all services, and employ concepts such as the circular economy and biomimicry. All this will enable them to manage themselves so as to regenerate both human and natural capital (see Sect. 3.8).

This arc of transition to what has been called *Natural Capitalism*[11] is starting to be implemented by a growing number of corporations. The 2016 UN Global Compact Accenture survey of more than 1000 CEOs found that 97% believed that sustainability is important to the future success of their business. Transparency was seen as critical factor, with 79% seeing brand, trust and reputation as driving action on sustainability.[12]

[8] Riane Eisler (2007).

[9] Introductory video, Humanistic Management Network, http://www.humanetwork.org/

[10] New Economics Foundation, http://www.happyplanetindex.org/

[11] Hawken et al. (1999).

[12] UN Global Compact 2016 – Accenture Strategy CEO Study, https://acnprod.accenture.com/us-en/insight-un-global-compact-ceo-study

Companies are starting to implement the first principle of Natural Capitalism: *use all resources dramatically more productively*. Neoclassical economists will tell you that markets make companies as efficient as they can cost effectively be. This is mostly mythology. Every company in the world can dramatically improve its resource productivity and thereby cut costs, unless the resources it uses are sold at very low, often subsidized prices. The entrepreneur and financier Jigar Shah suggests that approximately 50% of greenhouse gas emissions will be profitable to eliminate due to continuous technology innovation.[13] Such statements should not be misused, however, to declare carbon pricing unnecessary!

3.1.3 Redesign Everything

Significant structural transformations will be needed if an economy in service to life is to become a reality, and in many instances these are already underway. The second principle of Natural Capitalism is *to redesign how we deliver energy, feed ourselves, and make and deliver the services we desire, using such approaches as biomimicry and the Circular Economy.*

The discipline of Biomimicry, created by Janine Benyus, set forth principles on which nature does business. They are different from our current methods. Many corporations are now working with organizations, like the Biomimicry Guild,[14] to redesign how they make and deliver product and services using nature's principles. Nature makes a wide array of products and services, using only sunlight, with no long-lived toxins, at ambient temperature, using water based chemistry, wasting nothing. Companies that are implementing these approaches are again finding that it saves money and delivers superior service.

An auxiliary approach, outlined in Sect. 3.12.3, involves and requires action by the state: stop subsidizing the consumption of resources and instead make resources more expensive. In order not to damage the economy, it can be done in small steps and a revenue neutral manner keeping the overall tax level as low as it is, to reduce the fiscal load on the things we want for a finer future.

3.1.4 Regenerative Management

The third principle of Natural Capitalism is *to manage all institutions to be regenerative of human and natural capital.* The principles of a regenerative economy have been set forth in Club of Rome member John Fullerton's white paper,

[13] Shah, Jigar, Creating Climate Wealth, ICOSA 2013, http://creatingclimatewealth.co/

[14] http://biomimicry.net/about/

Regenerative Capitalism.[15] Like biomimicry, it draws from nature's principles, but applies them to running an economy in service to life.

Fullerton points out that there are patterns and principles that nature uses to build stable, healthy and sustainable systems throughout the world. These eight principles can guide us in creating an economy that operates in accordance with the rest of the world, creating conditions conducive to life:

1. *Right relationship*: Holding the continuation of life sacred and recognizing that the human economy is embedded in human culture, which is itself embedded in the biosphere.
2. *Innovative, adaptive and responsive*: Drawing on the innate ability of human beings to innovate and 'create anew' across all sectors of society.
3. *Views wealth holistically*: True wealth is defined in terms of the well-being of the 'whole', achieved through the harmonization of the multiple forms of capital.
4. *Empowered participation*: Financial wealth is equitably (although not necessarily equally) distributed in the context of an expanded view of true wealth.
5. *Robust circulatory flow*: A continual striving to minimize energy, material and resource throughput at all phases of the production cycle, reusing, remanufacturing and recycling materials.
6. *'Edge effect' abundance*: Creative collaborations increase the possibility of value-adding wealth creation through relationship, exchanges and resiliency.
7. *Seeks balance*: Balances resilience, the long run ability to learn and grow stronger from shocks with efficiency, which, while more dynamic, can create brittle concentrations of power.
8. *Honours community and place*: Operating to nurture healthy, stable communities and regions, both real and virtual, in a connected mosaic of place-centred economies.

These align with nature's underlying principles and are very similar to what we are learning from principles of human psychology and the emerging discipline of humanistic management.

Regenerative Capitalism is already manifesting in scalable, real world projects and enterprises on the ground. Capital Institute's Field Guide to investing in a regenerative economy profiles 34 companies that are implementing regenerative principles.[16] For the principles to become the 'source code' for the global economy, it will be necessary to apply them to large global enterprises.

Increasingly, big businesses are embracing the need for a new narrative. DNV-GL, a 150-year-old Norwegian company, is committing itself to strategies like creating a regenerative future. Owned by a trust, they are able to take a longer-term view of the company's responsibility than is possible in most listed companies. DNV-GL's Chief Sustainability Officer Bjørn Haugland states that a strategy for change 'should

[15] Fullerton, John, "Regenerative Capitalism," Capital Institute, 2015, http://capitalinstitute.org/wp-content/uploads/2015/04/2015-Regenerative-Capitalism-4-20-15-final.pdf

[16] Chang, Susan Arterian, "The Fieldguide to Investing in a Regenerative Economy," Capital Institute, http://fieldguide.capitalinstitute.org/

... speak to hearts as well as minds and inspire action and bring hope by communicating positive stories of change'.

These principles apply at least as much to the developing world. A great example is the work of Development Alternatives (Sect. 3.2).[17]

Similarly, the shift to regenerative agriculture can better feed people and at the same time suck carbon from the air and return it to the soil. Critics say that 'only conventional industrial agriculture can possibly feed people; we need GMOs, we need more artificial additives'. This turns out to be completely wrong (see Sect. 3.5). Indeed, the UN Food and Agriculture Organization (FAO) estimates that still 70% of all the food produced on earth comes from smallholder agriculture.[18]

This is good news: It means we do not have to remake most of agriculture, only help farmers who are still doing things mostly right avoid the mistakes of industrialized countries and gain access to the best regenerative practices (Fig. 3.1).[19]

The Savory Institute is involved in restoring the vast grasslands of the world through the teaching and practice of holistic management and holistic decision making. This has enabled practitioners to turn deserts into thriving grasslands, restore biodiversity, bring streams, rivers and water sources back to life, and combat poverty and hunger. Savory argues that this is the most promising way to deal with global climate change: mimicking how grasslands, the world's second largest carbon sink, co-evolved with massive herds of grazing animals. Holistically managed grazing animals are, he claims, one of the best ways to reclaim depleted land.[20] In nature, carbon is not the world's greatest poison.[21] Waste is a resource out of place. In nature, a use is found for it. Holistic management creates healthy communities of soil microorganisms to absorb carbon. Perhaps most importantly, it recarbonizes the soil and restores natural nitrogen cycles, in contrast to artificial carbon capture and storage, which has never worked at commercial scale and doubles the cost of coal plants.[22]

The importance of soils for carbon capture is one of the strong points of David Spratt and Philip Sutton.[23] More studies show that capturing one tonne of carbon per acre per year on average is reasonable on well-cared-for grasslands.[24] And Adam Sacks observes, 'We are only beginning to understand the potential of intensive

[17] Ashok Khosla, To Choose Our Future, 2016, https://www.amazon.com/Choose-Future-Paperback-Ashok-Khosla/dp/B01GMIAAUS

[18] Wolfensen (2013).

[19] Joel Salatin, Meet the Farmer, Parts 1–3, 29 April 2012, https://www.youtube.com/playlist?list=PL6C0D6709117A0049

[20] Savory Institute, Introduction to Savory Hubs, https://www.youtube.com/watch?v=SKWeqkq6tP4

[21] Allan Savory, How to Green the World's Deserts and Reverse Climate Change, TED, https://www.youtube.com/watch?v=vpTHi7O66pI

[22] Radford (2015), Also: Brown (2014).

[23] Spratt and Sutton (2008).

[24] A Landowner's Guide to Carbon Sequestration Credits, Central Minnesota Sustainable Development Partnership, P 8 http://www.cinram.umn.edu/publications/landowners_guide1.5-1.pdf

Fig. 3.1 Restoration success stories achieved by the Savory Institute

planned grazing with animals that break capped soil surfaces with their hooves, fertilize, moisturize and aerate the ground, and make earth hospitable to thousands of vital soil organisms. There is no climate-saving strategy that has anywhere near the potential of soils'. Sacks argues, 'There are roughly 12 billion acres worldwide, mostly ruined by human misuse, which we can restore. At a modest one-ton per acre we can pull twelve billion tons of carbon out of the atmosphere every year. That's six parts per million (ppm) – and even if we foolishly continue to add 2 ppm annually, it's still less than a 30-year trip back to a stable pre-industrial 280 ppm, down from today's perilous 393'.[25] We should mention that since Sacks wrote, the atmospheric CO_2 concentration has been measured at levels as high as 403 ppm.

Companies, communities and citizens are all recognizing that everyone's survival depends on behaving responsibly. Systemic policy changes are needed, including individual action, action by community groups and action by the corporate sector.

A Finer Future is Possible. It *is* possible for humanity to avoid total system collapse, and in so doing, create a finer future. Attaining it is the challenge for every human alive today.[26] Readers are invited to join.

3.2 Development Alternatives

Development Alternatives is an extremely encouraging example of an initiative taken in one of the poorest regions of the world that has created secure livelihoods, jobs, ecosystem health and optimistic perspectives for literally millions of people. The initiative was begun by Dr. Ashok Khosla who in 1982 left his comfortable career in government and the United Nations to set up a new type of institution designed to bridge the gaps between civil society and the government on the one hand, and civil society and business on the other. Ashok and his team were able to demonstrate that environmental problems are best addressed by dealing with their root causes. While in many cases immediate remedial measures are necessary, prevention through *alternative development strategies* can lead to cheaper, deeper and more long-lasting solutions, hence the name of the organization, *Development Alternatives* (DA).[27]

Development Alternatives was initiated with a $100,000 project grant from UNEP and was originally motivated and driven by an environmental mission. It started its operations by analysing what changes were needed in the existing systems of economy, society and governance to ensure that the health of the environment is maintained and regenerated for future generations. It recognized that since more than 70% of the people of India (and, indeed in much of the global South) live in villages and small towns, the environmental canvas must include the issues that concern them most, even if these are not commonly perceived by those in the urban or international community as being relevant to them.

The primary social goal of DA was to find and implement methods to empower people in a variety of ways. 'Empowerment', in its view, includes the ability to take part in the modern economy, the social standing to access entitlements and the authority to participate meaningfully in the institutions of family, community and local government. The end result of this chain of growing capacities is the creation

[25] Adam Sacks, "Putting Carbon Back in the Ground – The Way Nature Does It," http://www.climatecodered.org/2013/03/putting-carbon-back-into-ground-way.html

[26] Steffen (2015).

[27] Ashok Khosla emphasizes that much of DA's impact results from the efforts of its vast network of partners who, strengthened by its capacity building and other supports, greatly multiply activity on the ground.

of a citizenry that has more control over their decisions and their future. At its core, this means sustainable livelihoods – jobs or vocations that provide income, dignity and meaning. Such work should create goods and services for basic needs it must also conserve and rejuvenate the environment. This means that sustainable activities are both a basis for and the result of empowerment.

In line with the philosophy of Gandhi, technologies have to be more human in scale, less wasteful in terms of resources, and directly responsive to the basic needs of the people doing them. The possibility for such sustainable development is undermined if the economic and social disparities in a society are large. The extremely poor tend to overutilize and destroy (mainly the so-called renewable) resources due to the exigencies of survival and need. The very rich tend to overutilize and destroy other (mainly the non-renewable) ones, most often out of greed and entitlement. Increasing social equity, as well as eradicating poverty, thus becomes a primary instrument for environmental conservation.

To implement the concepts and designs it evolved, DA created a group of associated organizations that are formally independent, but have contractual obligations to put these concepts and designs into practice. In addition to the non-profit flagship Development Alternatives, they include the commercial wing, Technology and Action for Rural Advancement (TARA), and various subsidiaries that operate as business entities and are registered as companies which manufacture and market the technologies of Development Alternatives.

As a think tank, DA has done pioneering work in developing such concepts. Beyond these core goals, it has a highly innovative R&D facility that develops specific technologies that meet both criteria of environmental soundness and relevance to poverty eradication. These include machines and appliances for cooking (domestic woodstoves), electricity production (gasifiers), green, affordable construction materials (mud blocks, microconcrete tiles, ferrocement elements), weaving (advanced handlooms; see Fig. 3.2), handmade, recycled paper and other environmentally benign products for sustainable livelihoods. Much of DA's work has focused on regenerating the health of land, water and forest resources.

Fig. 3.2 The TARA 'Flying Shuttle Loom' – the powerloom without power: gives a weaver 3–4 times the income compared with a traditional loom (Photo: Development Alternatives)

Years of experience in the field proved to Ashok Khosla and his team that there can be no sustainable livelihoods if there is no ecological security – and vice versa: densely populated ecosystems cannot survive without sustainable livelihoods for all. Thus, in the constellation of interventions designed and put in place by Development Alternatives, sustainable livelihoods and environmental conservation are interchangeable concepts.

Can't the formal sector solve the unemployment problems?

The question is generally if the formal job sector can solve the unemployment problems.

According to the International Labour Organization (ILO), the world would need roughly one billion additional jobs to overcome global unemployment. This is actually Sustainable Development Goal (SDG). It would mean that developing countries need to create more than 50 million new jobs every year. At the same time, even in the poorest countries, the capacity of agriculture to absorb additional labour is rapidly diminishing.

Box: Livelihoods Cost Little Compared to Conventional Jobs

According to mainstream economics, the creation of livelihoods and jobs should, generally, be the job of the corporate sector. However, the corporate sector is not currently set up to creating jobs or livelihoods in Third World economies in the numbers needed. The compulsions of global competition encourage industrialists to invest in machines rather than in people. Global competition dictates criteria leading to high cost per new job created. The capital required to create one industrial workplace in an industrialized country is about $1000,000. Even if wages are much lower in developing countries, new industrial jobs can also be very expensive there.

Such heavy initial investments act as a major barrier to the creation of new enterprises in the formal sector and therefore to jobs.

Even India, which is seen as a success story in terms of creating new manufacturing jobs, has not been able to enjoy a net increase of such employment. During the past 25 years, the economy has reached new heights but the total number of persons employed in the large, formal corporate sector in India (including the Business Process Outsourcing - BPO) has actually remained where it was.

So it seems extremely unlikely that SDG 8 will be achieved by additional jobs in the formal sector.

The consideration in the above box leaves us with the need that somebody else will have to take responsibility for creating jobs and sustainable livelihoods. This is where the small and medium enterprise sector comes in: market-based, profit-making businesses that are mostly small and generally local. In most economies, they are the largest generators of jobs and livelihoods. Development Alternatives is an innovative part of this segment.

Perhaps the most important yet least understood impact of large-scale livelihood creation is on a nation's demography. Together with programmes for the education of girls and women, sustainable livelihoods are probably the most effective stimuli for smaller families and lower birth rates. For the longer-term interests of planetary health, it is in the interest of all, rich and poor, to accelerate the process by which the demographic transition to low fertility and, as a result, to low population growth is achieved in poorer countries. And lower fertility rates can be the most powerful answer to the unemployment challenge!

Over its 30-year lifespan, Development Alternatives has implemented some 700 projects at a cost of more than US$ 150 million. These have led to the creation of large-scale changes in the lives of people, particularly in remote communities mainly in north and central India. These changes include the creation of jobs and livelihoods as the end products of many intermediate steps that can be summed up, as mentioned earlier, by the term 'empowerment'. It is not difficult to estimate the 'direct' jobs generated by the industries created using its technologies, or the number of farmers who immediately gained jobs because of its check dams and water management systems, or the number of women who were able to get paid work because of the time saved from no longer having to collect water or fuel wood. However, given the systemic nature of the approach of Development Alternatives, these economic activities lead to second and third level jobs that are the downstream multipliers of the direct jobs – and these are substantially greater in numbers that are not easy to calculate.

Many of the people who are 'empowered' – through acquisition of skills, assets, entitlements, access to knowledge of work opportunities and other similar factors – are enabled to get jobs or create livelihoods for themselves. The portion of those empowered in these different ways who actually become gainfully employed varies (based on field experience) between 10% and 30%. For women, this percentage is probably lower because of their limited mobility in traditional societies.

Development Alternatives generally takes the lower figure to estimate its impacts on livelihood and job creation, that is, around 1.6 million, in addition to the direct and indirect jobs created, which are estimated to be in the region of 1 million. So external observers estimate that Development Alternatives has been responsible, directly and indirectly, for the creation of about 2.6 million jobs over the 30 odd years of its existence. A fabulous record, everybody would agree. However, if the downstream multiplier jobs are also counted, the ultimate figure could well be in the vicinity of 5 million.

One important element for the poor to gain livelihoods is education. In this context, DA has developed the 'Literacy to Self-Reliance' programme, an end-to-end solution especially catering to the basic literacy and numeracy needs of rural women. As a first step, the enrolled women are provided with functional literacy through the ICT-based programme 'TARA Akshar+', a computer-based teaching programme that uses advanced teaching techniques, enabling a learner to read and write in Hindi, as well as simple math calculations, within 56 days and at very low cost. In a second step, post-basic learning is taken on by the TARA Livelihood Academy, the skill development arm of DA. Over a period of 2 months, those enrolled can acquire the vocational and business skills which enables many of these women to become entrepreneurs and leaders in their communities.

TARA, the commercial wing of the DA Group, incubates enterprise ideas and business models for take-up by aspiring micro and small entrepreneurs. TARA aims at promoting low-carbon pathways and inclusive growth through enterprise development in the sectors of rural housing, renewable energy, water management, sustainable agriculture and waste management and recycling. Together with its partners, TARA has facilitated the establishment of more than a thousand enterprises since its inception, thus having mobilized local economies and created green jobs. One arm of TARA is the TARA Machines and Tech Services Pvt. Ltd., a social enterprise that has developed small 'waste to wealth' business packages based on green building material production technologies. Their product range is developed for the Indian market to cater specifically to small-scale enterprises that in turn make affordable, green construction products available in rural India.

Another activity of DA has been the building of more than 400 'check dams' which counteract the alarming trend in India of falling groundwater levels. The chronic drought situation in Bundelkhand has aggravated over the years. But the check dams were the right answer to the calamity. They have revolutionized water security at very little cost. Essentially, they keep surface water running slowly, which also leads to a much better use of the water for agricultural crops. The increase of productivity is about 25%, leading to much higher income for the local farmers. Impressed by DA's success in the 1980s, the check-dam technology has been widely adopted by other development organizations and the Indian Government for large-scale implementation across the country.

Check Dam:

200 Ha Irrigated
2 more Crops per year

Groundwater Recharged

200 Livelihood
Cost: 8000 €. ROI: 200+%

Development Alternatives has become one of the first major international NGOs headquartered in India. DA and its associated production and marketing organization TARA are now the premier institutions in India in the field of environmentally sustainable development with more than 800 staff operating in various parts of India.

The following chart provides an impressive picture that summarizes DA's success story over three decades on the empowerment of people (Fig. 3.3).

Similar pictures exist for environmental and job achievements such as CO_2 emissions reduction by 850,000 tonnes and water consumption reduction by 935 million litres. For the jobs, the modest figure mentioned before of 2.6 million is valid.

The DA Group defines a transformative agenda for India as the restructuring of institutional systems as well as a shift of virtually all attitudes towards society and

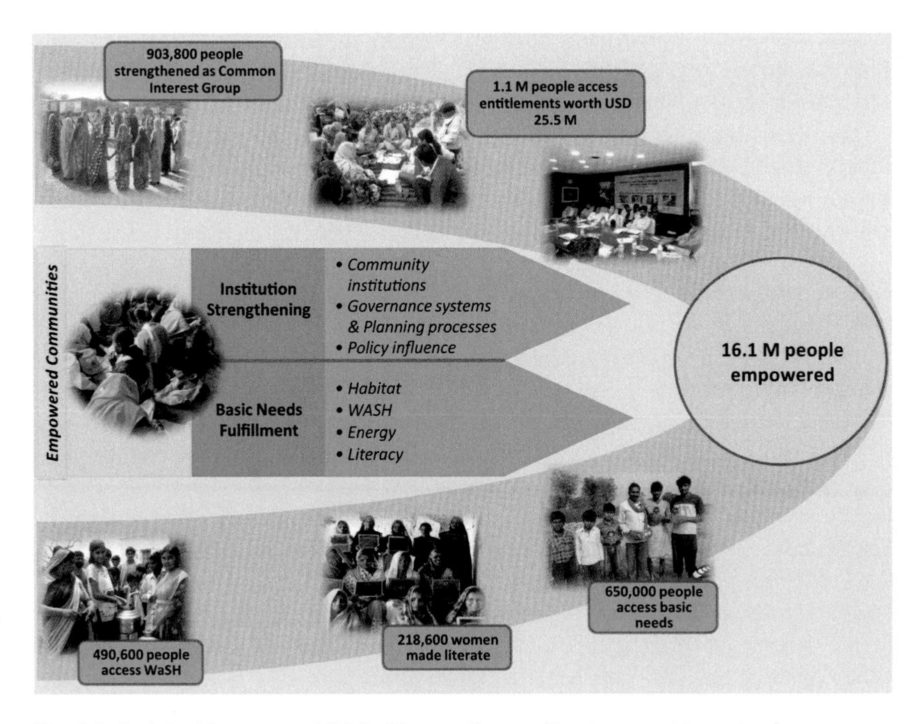

Fig. 3.3 Social achievements of DA in 30 years (Source: Development Alternatives)

economy – encompassing consumption and production patterns, well-being and justice and enterprises and governance. Its core is the redefinition of the economy as a *subsystem* of society and nature and thus as a means towards a socially just and environmentally sustainable future – not an end in itself, as it currently stands. Instead of following the development path of Western societies, going through a phase of intense industrialization with highly negative environmental and social impacts, DA stresses the need for India to define its own development path – at best taking a shortcut from its current situation directly onto a sustainable trajectory.

3.3 The Blue Economy

In April 2009, Gunter Pauli presented some core research and concepts on local economic development as a possible report to the Club of Rome, under the title 'The Blue Economy: 100 innovations, 10 years, 100 million jobs'.[28] He was sketching out a bold vision inspired by the German proverb *Schaffen ist auch Wissenschaft* (Doing is also Science). This vision was based on the understanding that nature in general (and a wide range of ecosystems in particular) has overcome nearly every imaginable challenge over the millennia. Nature, therefore, provides the inspiration

[28] Pauli (2010, 2015)

for how human society can chart a pathway towards the future. The pathway may be derived from the ingenuity of ecosystems that continue to provide a wealth of products and services on which all life depends.

This new 'Blue Economy' would strengthen all the social systems that build up culture, tradition and social capital, because they provide resilience in adverse times and joy in good ones. It would also permit us to learn how to live within obvious limits as we evolve from scarcity to abundance. The Club of Rome's Executive Committee enthusiastically encouraged Pauli to pursue this line of thinking and write and present the book. It became a huge success and so far has been translated into 41 languages.

Pauli had been observing ecological and social systems for decades, which enabled him to postulate a few core principles (see the box below) that could guide the quest towards a world where nature regains its evolutionary path and society strengthens its social web. This goal would enhance the quality of life for everyone by empowering them with the knowledge of how to fulfil basic needs with what is locally available. The book was published in 2010, and the years after have taught Pauli many new lessons. The original vision and the 100 proposed innovations have been tested for practicality. He adapted the guiding principles as an effort to explain how fast shifts can be achieved from today's mainstream business philosophy based on the logic of globalization, cost cutting and ever higher economies of scale, to a Blue Economy that performs better and transforms industries faster than often has been considered viable. It all has to start with our capacity to respond to the basic needs of all.

The quest for food security goes hand in hand with the need to produce and consume within our planetary boundaries. There is an urgent need to respond to the basic needs of all on Earth, while there is an equal need to shift towards healthier nutrition. The combination of food security, sustainable farming and health concerns is forcing the world to embrace innovations. These will be social, technological and organizational. It is clear that one technology will not offer a complete solution. Any response to the challenges faced today will require a cluster of responses that will evolve over time. More of the same will rarely produce better results. But some basic principles can guide us in our search for breakthroughs.

Box: 21 Principles of the Blue Economy (2016 Edition)

1. Product and consumption systems inspired by nature
2. These systems are *non-linear*
3. Systems *optimize* (not maximize) and *co-evolve*
4. Systems demonstrate *resilience* through an ever-increasing *diversity*
5. Systems operate first in the basis of *physics*, the adapt chemistry and biology
6. Products are renewable first, always organic and biodegradable
7. Success in performance depends on a *change the rules of the game*
8. *Isolated problems* are interconnected to create *portfolio of opportunities*

9. Performance include the power to *put nature back on its evolutionary and symbiotic path*
10. The multiple benefits include the *strengthening of the Commons*
11. The purpose is to first respond to the basic needs
12. Use what you have
13. *Replace something with nothing*, eliminating unnecessary products
14. Everything has *value*, even waste and weeds
15. *Health and happiness* is the result
16. *Economies of scope*, producing in clusters instead of economies of scale
17. One Initiative generates *multiple cash flows and multiple benefits*
18. *Vertical integration* of the *value chain* in both *primary* and *secondary* industries
19. *Management free of business plans*, but driven by complex systems analyses
20. All decisions have an impact on *profits and losses and balance sheet*
21. All ethics have *ethics at the core*

3.3.1 Core Principles

In 1994, Gunter Pauli, then working for the United Nations University, UNU in Tokyo, founded the Zero Emissions Research Initiative (ZERI) and developed a network of scientists who collectively think beyond the obvious. Their search for solutions originally to be presented at the Third Conference of the Parties of the Climate Convention ('COP3') in Kyoto in 1997 was all inspired by 'how nature evolves from scarcity to abundance'. It started from the observation that the only species on Earth capable of producing something no one desires is the human species. Nature continuously cascades matter, energy and nutrition, and every member contributes to the best of its capabilities. The concept of unemployment does not exist in ecosystems. It is against this idealistic framework that ZERI engaged in the design of business models that increase resource efficiency while generating more food and nutrition than ever imagined before, solely using locally available resources. The updated core principles are represented in the box above.

A small selection of its practical innovations may serve as the paragon of the optimistic 'Come On' slogan.

3.3.2 Coffee Chemistry and Edible Mushrooms

Agricultural programmes relying on manipulating genes often ignore that our present industrial farming and food production model is extremely wasteful. Do we realize that a cup of coffee only contains 0.2% of the biomass of the red (coffee) cherries harvested? The process of fermenting, drying, roasting, grinding and

Fig. 3.4 Mushrooms growing on coffee plantation biomass. One out of 200 examples for cascade use of natural resources in the Blue Economy (Photo: Development Alternatives)

brewing leads to the ingestion of a minute fraction of the ten million tonnes of coffee produced worldwide, and the disposal as waste of nearly everything else.

This understanding has given rise to the 'coffee chemistry', including the farming of mushrooms on post-harvest, post-industrial and post-consumer coffee biomass, the use of the spent substrate enriched with amino acids as animal feed, the use of fine coffee particles as an odour control, UV protector and even hydrogen storage system. The logic of coffee can be applied to tea and dozens of other crops. This bundling of innovations not only allows for the substitution of toxic chemicals, it also generates income and jobs. Figure 3.4 shows a rich, ready-to-harvest crop of mushrooms thriving on coffee plant biomass.

3.3.3 Designing Biorefineries and Thistles in Sardinia

Recent cases demonstrate that the generation of 500 times more nutrition, from the same coffee harvest, and the creation of 300 times more value from readily available biomass is not an exception. Over the past 20 years, the partners of the ZERI Foundation have discovered dozens of other cases which are now scaling up as demonstrated by more than 5000 farms that combine coffee and mushrooms. The design of biorefineries offers more insights into the dynamics of food and chemistry, with the critical success factor being the availability of feedstock.

The case of Novamont in Sardinia demonstrates that the processing of thistles, a weed that grows prolifically on abandoned farmland, can respond to multiple needs in society, while offering a new perspective to farming. Thistles are harvested, processed as oil or sugars from cellulose, then converted into a portfolio of biochemicals including polymers for plastic bags, elastomers for rubber gloves, herbicides and lubricants, and the waste can be processed into animal feed.[29]

[29] https://ec.europa.eu/eip/agriculture/en/content/eip-agri-workshop-building-new-biomass-supply-chains-bio-based-economy

3.3.4 3D Sea Farming and Fishing with Air Bubbles

The portfolio of innovative business models is not limited to farming on land. The introduction of 3D sea farming, combining the cultivation of seaweeds, mussels, scallops, oysters, fish, crabs and lobsters has demonstrated to be a highly effective way of reviving the healthy production of seafood. In a controlled environment, it secures a diverse output ranging from food and animal feed to ingredients for cosmetics and pharmaceuticals, with any remaining waste converted to fertilizers. This system requires no inputs like fresh water, pesticides or fertilizers; on the contrary, this technique – which is considered the permaculture of the sea – alkalizes the seawater, regenerates biodiversity and helps shift the diet of consumers towards a much healthier one.

One of the most profound changes in modern food production systems is fishing and fish farming. The age of nets, hooks and cages are over. The conversion of sardines to salmon feed is absurd when we need to double food output. The ZERI Foundation, inspired by the way dolphins and whales catch their prey, has focused on the design of fishing techniques that rely on air bubbles, and this has led to a redesign of fishing vessels and techniques. All female fishes with eggs are released into the ocean to secure future generations and provide an ample supply of wild catch. Indeed, one reason that fish farming is perceived as more productive than fishing is because fishermen indiscriminately kill females with eggs, while farm can keep them alive.

Time has come to innovate and encourage innovators to disrupt the old, destructive practices. This can not only generate jobs, but can also turn the old productivity logic on its head. A 3D fish farm can generate two jobs per hectare on certain areas of the sea requiring 25 lines at a total cost of $7500 generating 600,000 shellfish and 75 tonnes of seaweed per hectare per year, according to Gunter Pauli. All conform with the philosophy of the Blue Economy: more value, less investment costs, more output and jobs. And humans may even get healthier in the process.[30]

3.4 Decentralized Energy

Amory Lovins and his team at the Rocky Mountain Institute in *Reinventing Fire* have developed a great vision: 'Imagine fuel without fear. No climate change. No oil spills, dead coal miners, dirty air, devastated lands, lost wildlife. No energy poverty. No oil-fed wars, tyrannies, or terrorists. Nothing to run out. Nothing to cut off. Nothing to worry about. Just energy abundance, benign and affordable, for all, for ever'.[31]

[30] For reference on this and other Blue Economy examples go to webpages www.zeri.org and www.TheBlueEconomy.org. Readers can also get in touch by Twitter @MyBlueEconomy.

[31] Lovins and Rocky Mountain Institute (2011).

Well, this is the wonderful visionary language of pioneers like Amory Lovins. Of course, *No climate change* is unlikely to become a reality any time soon. *No lost wildlife* is euphemism too: much of the current trends related to the expansion of renewable energies are eating up a lot of land that could otherwise be left for wildlife. And the term *energy abundance* sounds like an invitation to wasteful use – which is the opposite of the book's message. The significance of *Reinventing Fire*, however, does not lie in its minor language exaggerations but in its major truth.

At the time of its writing, 6 years ago, only very few people foresaw the profound changes the energy industry would soon be undergoing. Now they are on everybody's mind. Entire countries experience massive changes, mostly to a more sustainable energy future, that in a general sense run along the lines of Amory Lovins' vision. Classical centralized power utilities have come under severe stress from competing renewable energies. The energy industry is currently undergoing massive changes, and most are moving towards a more sustainable future.

Denmark and Germany were the trendsetters. Denmark in 1985 adopted a law that prohibited the installation of nuclear power utilities and instead promoted wind energy. That was a year before the Chernobyl nuclear disaster! Germany began talking about a phase out of nuclear energy just after Chernobyl and adopted a phase out law in 1999. At nearly the same time, it introduced a generous feed-in tariff (FIT) law for renewable power, leading to a stunning boom in renewables. After the Fukushima nuclear tragedy, Germany accelerated the phase out while enjoying rapid progress of renewable energies.

China (and some 60 other countries) more or less copied Germany's FIT schemes, leading to a rapid development of technical innovations and economies of scale. Figure 3.5 shows how prices for solar photovoltaics (PV) tumbled while cost of nuclear power soared. Since 2010, there is no economic argument left for investing in nuclear power.

Globally, renewables have been increasing dramatically. In Chile recently, so much solar energy was produced that the utility gave it away for free.[32] Germany has pledged to be 100% renewable by 2050, Scotland by 2020. According to the Asia Europe Clean Energy Advisory (AECEA), China has installed 34.2 GW of solar energy in the year 2016 alone.

Some analysts peg the tipping point as 2014, when renewable energy began what is now increasingly seen as the inevitable triumph of wind and sun. In April 2015, Michael Liebreich of Bloomberg New Energy announced, 'Fossil fuel just lost the race with renewables….The world is now adding more capacity for renewable power each year than coal, natural gas, and oil combined'.[33]

Coal's competition drop is clearly illustrated on the stock market. Figure 3.6 shows the stock market performance of average US stocks (Dow Jones Index) compared with the stocks of coal power companies.

In the long run, it is clear that the world's economy, instead of being powered by fossil and nuclear fuels eventually will be based on renewable energy sources. The

[32] Dezem and Quiroga (2016).

[33] Randall (2015).

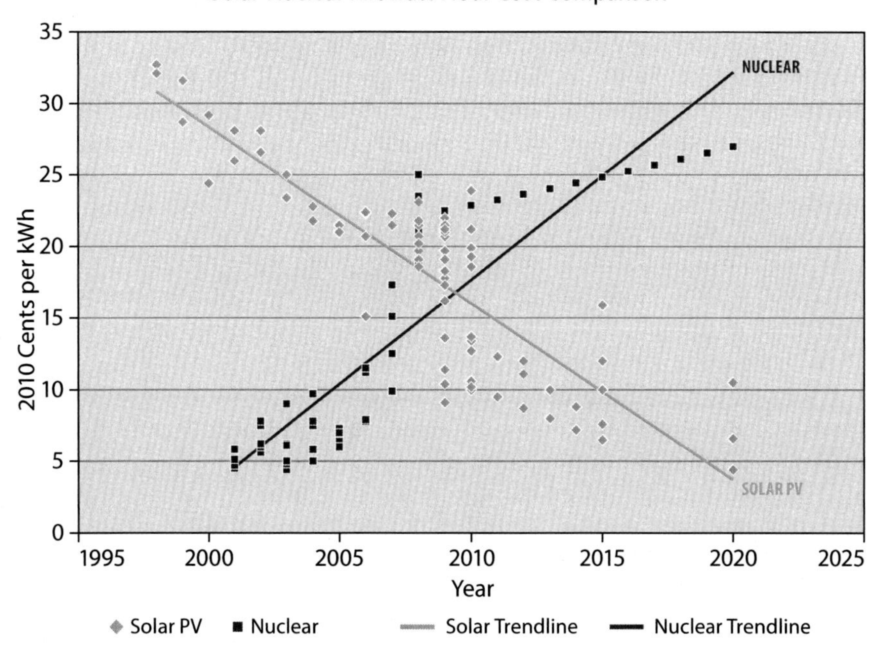

Fig. 3.5 Solar PV beats nuclear in terms of cost (Source: NC WARN)

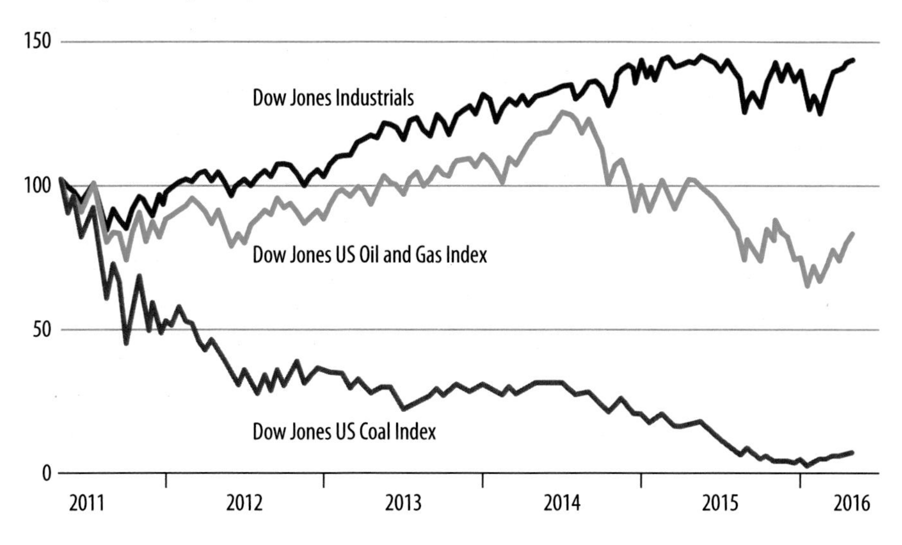

Fig. 3.6 Over 5 years, coal stocks lost most of their value, while the Dow Jones Index was up >40% (Source: TruValue Labs, June 29, 2016)

crucial issue for the planet's future is how long it takes for this transition to be fully implemented. Fossil and uranium/thorium resource depletion is not the chief reason. It's the political, ecological and technological cost of failing to react to global warming and to increasing costs of the entire nuclear cycle.[34]

This sounds as if the transition to a renewable energy world is not only necessary but imminent extremely soon. However, leaving large fossil fuel reserves in the ground has a corollary. There will be 'stranded carbon assets', that is, former economic assets becoming valueless or even a liability. A recent study[35] estimates these 'stranded assets to be around $ 6 trillion, assuming that only 20% of identified fossil fuel reserves can be burned by 2050'. Other guesses even arrive at $20 trillion, resulting from a cap on carbon emissions designed to limit warming to 2 °C.[36] These 'stranded assets' can be considered as part of the cost of making the transition of phasing out fossil fuels.

A slightly parallel consideration applies to the phase-out of nuclear energy. Here it is politically more complicated because several countries have since decades entertained a synergetic economy between military and civilian nuclear power, in other words, have tacitly subsidized civilian nuclear power.

Given the stranded assets problem, it is interesting to ask how fast the world can achieve the transition to an economy based only on renewable energy resources. This question has been studied recently by Mark Z. Jacobsen and his colleagues at Stanford and UC Berkeley. They argue that it is possible to achieve a total transition of the world out of fossil fuels by 2050.[37] Figure 3.7 shows how the transition from fossil fuel dominance to a fully WWS (wind-water-solar) powered world is achieved in their study.

A number of comments should be made regarding the figure. First, when aggregating across the world, one should keep in mind that the percentage ratio of solar power to wind power varies significantly in different countries. Second, note from Fig. 3.7 that the end-use power load in 2050 is slightly lower than what it was in 2012. However, this load is *much less* than the end-use power, which would be needed in a business-as-usual scenario without a WWS transition. This means that the authors assume a considerable potential for efficiency gains. Third, conversely, it has to be admitted that the WWS scenario requires more than solar panels, wind parks, etc. The issues of managing peak demand and power intermittency also have

---

[34] Gilding (2015).

[35] Carbon tracker and the Grantham Research Institute on Climate Change and the Environment at the LSE. http://www.carbontracker.org/report/unburnable-carbon-wasted-capital-and-stranded-assets/

[36] Capital Institute. July 19, 2011. The Big Choice. In The Future of Finance Blog, also referring to Carbon Tracker.

[37] Mark Z. Jacobson, Mark A. Delucchi, Zack A.F. Bauer, Savannah C. Goodman, William E. Chapman, Mary A. Cameron, Cedric Bozonnat, Liat Chobadi, Jenny R. Erwin, Simone N. Fobi, Owen K. Goldstrom, Sophie H. Harrison, Ted M. Kwasnik, Jonathan Lo, Jingyi Liu, Chun J. Yi, Sean B. Morris, Kevin R. Moy1, Patrick L. O'Neill, Stephanie Redfern, Robin Schucker, Mike A. Sontag, Jingfan Wang, Eric Weiner, Alex S. Yachanin Draft paper December 13 2015, 100% Clean and Renewable Wind, Water, and Sunlight (WWS) All Sector Energy Roadmaps for 139 Countries of the World, http://web.stanford.edu/group/efmh/jacobson/Articles/I/CountriesWWS.pdf

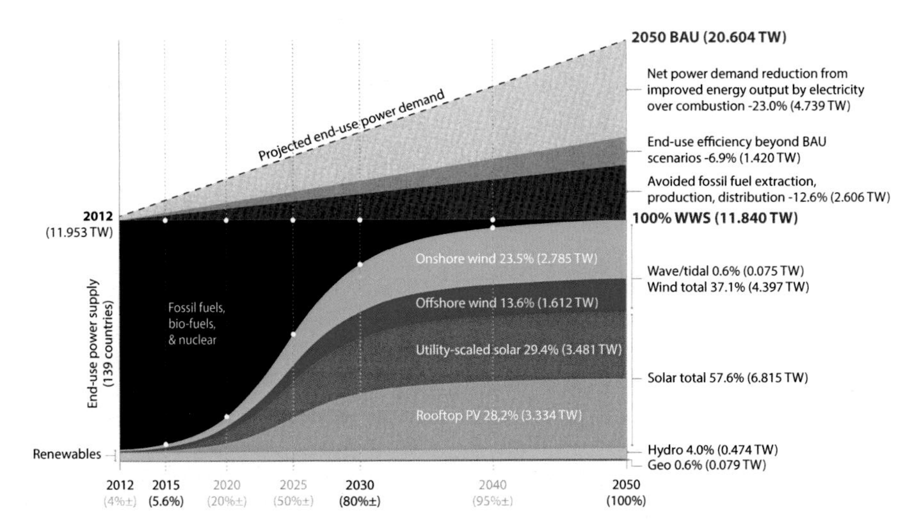

Fig. 3.7 Aggregate changes for 139-country end-use power demand for all purposes and its supply by conventional fuels and WWS generators over time (Source: see Jacobson et al., footnote 6)

to be handled and financed, but both are solvable problems, and inexpensive compared to the side effects of fossil fuels.

Let us for a moment turn to the political side. In principle, two steps are essential to accelerate the transition from a fossil fuel-driven economy to one being powered by renewable energy:

1. All financial incentives to the fossil fuel industry (both privately owned and state owned) should be eliminated. A recent IMF paper[38] estimates these so-called pre-tax subsidies globally to the fossil fuel industry to be in the range of about $600 billion annually.
2. Impose an internationally harmonized, but nationally retained, carbon tax (see Sect. 3.7.3). For most developing countries, these trends towards an increasing dependency on renewable energies will come as a blessing. Decentralized energy use is technically feasible and can greatly help create jobs where they are most needed, in rural areas of developing countries.

In the long run, incumbent electric utility and coal and oil companies will either join the transition[39] or find themselves out of the job.[40] Early in 2016, the consulting company Deloitte predicted that more than 35% of independent oil companies would go bankrupt, with another 30% to follow in 2017. This on the heels of 50 North American extractors bankrupted in late 2015.[41] Surely, some newly increasing oil prices, and the intended fossil fuels protectionism of the Trump Administration

---

[38] Coady et al. (2015).

[39] Uphadhyay (2016).

[40] Ahmed (2016).

[41] Deloitte Center for Energy Solutions (2016), See also Arent (2016) (from Goldman Sachs).

may change the picture, at least temporarily. The stranded assets crisis can actually be considered the strongest motive for Donald Trump's otherwise ridiculous preference for fossil fuels.

The biggest user of energy, China, is becoming the world's renewable energy powerhouse.[42] Growing its installed solar capacity by 20-fold within only 4 years, China went from a capacity of 0.3 GW in 2009 to 13 GW by 2013, adding 30.5 GW of renewable energy in 2015, 16.5 GW of that solar. China still burns a lot of coal, but its Green Horizons programme has committed to clean the air in its cities and cut carbon intensity by 40–45% from 2005 levels within the next 5 years.[43] In 2016, the total recorded CO_2 emissions were reduced by 5% despite a 7% economic growth rate. And by 2050, China intends to get 80% of its energy demand from renewable energies.[44]

China has announced that it intends to become the 'Ecological Civilization', a concept that it wrote into its constitution in 2012. China's 13th Five-Year Plan (Sect. 3.16.1) is impressively reflecting this new mind-set.

A fortunate side effect is that the global switch to renewable energies – and indeed energy efficiency – is accompanied by an increase of jobs. The International Renewable Energy Agency, IRENA, recently noted that renewable jobs are growing at 5% a years, now exceeding eight million globally. These jobs are more dominantly in manufacturing than jobs in general, and tend to deliver greater gender parity.[45]

There are assessments that the whole transition could come even faster. Stanford Professor Tony Seba predicts that by 2030 the entire world will be using renewable energy – not just electricity, but all forms of energy. Seba's book, *Clean Disruption*,[46] describes why he believes the transformation will come so fast. He credits four factors: the fall in the cost of solar, the fall in the cost of storage (batteries), the electric car and the driverless car. With transportation responsible for 30% of carbon pollution, displacing oil-based transport will be as big a deal as bankrupting coal. Seba uses the analogy of how the experts totally underestimated sales of mobile phones. In the 1990s, McKinsey told AT&T that it could only expect 900,000 mobile subscribers, but the real figure turned out to be above 108 million.[47] Seba asks: Don't believe in the Clean Disruption? The IEA wants you to invest $40 trillion in conven-

---

[42] IBM Research Launches Project "Green Horizon" to Help China Deliver on Ambitious Energy and Environmental Goals, 7 Jan 2014, http://www-03.ibm.com/press/us/en/pressrelease/44202.wss

[43] "CHINA TO APPROVE OVER 17.8GW OF PV IN 2015, Bloomberg New Energy Finance, http://about.bnef.com/landing-pages/china-approve-17-8gw-pv-2015/

[44] "China 2050 High Renewable Energy Penetration Scenario and Roadmap Study" China National Renewable Energy Centre, http://www.rff.org/Documents/Events/150420-Zhongying-ChinaEnergyRoadmap-Slides.pdf

[45] "Renewable Energy and Jobs," International Renewable Energy Agency 2016, http://www.irena.org/DocumentDownloads/Publications/IRENA_RE_Jobs_Annual_Review_2016.pdf

[46] Seba (2014).

[47] Kim (2013).

tional energy (nuclear, oil, gas, coal) and conventional utilities. It's their Kodak moment. It's your money.[48]

Entrepreneur Elon Musk created a car company with a market capitalization well more than half of General Motors, despite selling 300 times fewer cars?[49] How can that be? Because Tesla, as its Master Plan Part Deux, released in July 2016, makes clear, is integrating rooftop solar with home battery storage and with electric cars. Tesla is really a battery company, and if battery costs come down as they have been doing, then the game really is over for the fossils. By integrating energy generation with storage and with transportation, Tesla may eliminate pretty much any rationale to dig up and burn ancient sunlight in ways that are dirty and dangerous, and politically destabilizing.

3.5 Some Agricultural Success Stories

3.5.1 General Lines of Sustainable Agriculture Policy

A separate book would be needed to deal with benign and ecologically sustainable agriculture. In Sect. 1.8, we said that the currently dominant global agriculture is in no way sustainable and took the IAASTD Report as a better alternative. The main policy findings of the report emphasize the *multifunctionality* of agriculture: as a provider of food, social security, ecosystem services, landscape value and more.[50] This contrasts with the agro-industry approach, which is divorced from social and ecological features and focussed exclusively on maximum production.

The practice of sustainable farming has been termed agro-ecology, covering a wide range of systems that are adapted to local conditions and refined to meet local needs. Common to all these is the principle of ecological, economic and social sustainability. Agro-ecology preserves soils and water supplies, regenerates and retains natural soil fertility and encourages biodiversity; its yields are sustainable in the longer term. To a large extent, it avoids agrochemicals by growing diverse crops together and copies nature's closed material flows. It sequesters carbon rather than adding to emissions. At the same time, it allows farmers to earn enough money to live on; it develops processing facilities to protect jobs in rural areas while paying farmers a fair reward for their produce and a reasonable recompense for their work to protect nature and the climate.

---

[48] Tony Seba, 2012 How to Lose $40 Trillion, Tony Seba, http://tonyseba.com/how-to-lose-40-trillion/ By 2015 the IEA was saying that the number is $48 trillion by 2035. Either way it is a huge sum, which the IEA proposes be spend essentially entirely on oil, gas, coal and nuclear. https://www.iea.org/newsroomandevents/pressreleases/2014/june/world-needs-48-trillion-in-investment-to-meet-its-energy-needs-to-2035.html

[49] Ogg (2016).

[50] See also UNEP and International Resource Panel (2014).

Policies to support these aims could be adopted by governments of both North and South and may include (1) providing secure access to land, water, seeds, information, credit and markets; (2) revising laws of ownership to support women, farmers and indigenous and community-based organizations; (3) establishing more equitable regional and global trade arrangements; (4) revising intellectual property laws to respect farmers' rights and address equity goals and biodiversity; (5) investing in local infrastructure and agro-processing; and (6) increasing public research and extension investment.

A recent report by UNEP's International Resource Panel (IRP)[51] supports the IAASTD critique of the current food systems, saying they are responsible for 60% of global terrestrial biodiversity loss and about 24% of global greenhouse gas emissions. The IRP proposes a 'resource-smart' food system, based on three principles: low environmental impacts, the sustainable use of renewable resources and the efficient use of all resources. The IRP report suggests that resource efficiency gains of up to 30% are attainable!

Yet, instead of a book-size description of a sustainable system of agriculture and food, we opt for a modest and brief description of a few examples.

3.5.2 Sustainable Farming in the Developing World

Some practical examples can illustrate the general lines of a sustainable agriculture and food system that helps both farmers and consumers.

The global cocoa market is highly monopolized – 80% of production is controlled by two transnational companies. Most of the cocoa they buy is produced in West Africa and it is virtually impossible to trace the cocoa back and identify the social and ecological conditions of production. The German chocolate producer *Ritter Sport* was dissatisfied with this situation and decided to support more sustainable standards in cocoa production, adopting the ambitious goal that 100% of their cocoa must be produced sustainably by 2018.

Ritter Sport has provided Nicaraguan small-scale farmers with education and training programmes since the 1990s and in 2001 assisted the founding of Nicaragua's first cocoa cooperative called 'Cacaonica', a term composed of cocoa and Nicaragua. Over 15 years, this initiative has developed into a collaboration with more than 3500 farmers, now organized into more than 20 cooperatives. This enterprise is based on agroforestry systems as an ecologically sustainable alternative to cocoa monocultures. The mix of different plant species allows farmers to achieve higher quality cocoa and increased income. The German partner adopted a payment model combining quality-based surcharges on the global market price and fixed purchase quantities. These help farmers to plan ahead and to secure their futures.

[51] UNEP (2016), Food Systems and Natural Resources (2016).

Together with the French chocolate producer CEMOI, Ritter Sport has now set up a similar educational and trade model in Ivory Coast in Africa. The company sees its initiatives as trade partnerships between equals rather than a form of development aid.

Another example of successful cooperative farming is found in Cuba.[52] When the Soviet Union collapsed, the sudden withdrawal of food and oil support precipitated an agricultural crisis. Almost overnight, fuels, trucks, farm machinery, spare parts and fertilizers and pesticides became very scarce. Over 40% of state farmland was restructured into 2000 new cooperatives managed by the workers, who were also allotted a gardening space to grow their own family's food. By 2000, more than 190,000 urban residents had also claimed personal lots on vacant city land[53]: small scale urban production minimized the need for oil for transportation and agricultural machinery. The absence of agrochemicals necessitated agro-ecological production. The so-called *organoponicos*, rectangular-walled constructions of raised beds containing a mixture of soil and organic material such as compost, became the mainstays of vegetable cultivation in Cuban urban agriculture.

Production increased enormously, and in Cuba's cities, the environment benefited from crop cultivation, often coupled with urban reforestation and ecological farming methods. Programmes to manage organic manures, seeds, irrigation and drainage, marketing and technical education supported crops and animal husbandry. Over 350,000 new, well-paying and productive jobs were created in these programmes over 12 years.

Another successful mode of ecological farming is the 'System of Rice Intensification' (SRI). It aims to mobilize biological processes that are already present in crop plants and in the soil that supports them. It was started in Madagascar in the 1980s by a Jesuit priest and local farmers who identified practices that improved production in irrigated rice paddy. It was gradually extended to rain-fed rice and many other grains and vegetables.

SRI is a work in progress, 'not a material set of inputs… to be implemented like the Green Revolution technology [but]… a set of ideas or insights'.[54] As such, it is rarely adopted by market-oriented players who aim to sell products. The ideas have been spread by civil society mechanisms, and have now reached many developing countries.[55]

Among SRI's innovations are planting seedlings rather than broadcasting seed; spacing plants; stimulation of roots mainly by building and maintaining natural fertility with decomposed organic materials that encourage soil organisms; avoiding waterlogging, especially helpful with irrigated rice, where intermittent flooding was found to produce higher yields than traditional methods; and aerating soils throughout the growth period.

[52] Koont (2009).

[53] Higgs (2014, pp. 12–13).

[54] SRI-Rice (2014).

[55] Uphoff (2008).

The proven advantages have included: improved yields (sometimes double or more) achieved without relying on improved varieties or agrochemical inputs; less need for external inputs in general, including savings on water and seed; and carbon sequestration.

Another novel approach to crop management exploits the natural relationships between plants and insects. When scientists investigated the ecology of a widespread cereal pest in eastern Africa, they discovered that adding selected forage plants into maize fields dramatically improved cereal yields and total farm output. The so-called 'push-pull' technology that emerged from their research uses natural plant chemicals that drive insect pests away from the intended crop and attract them to other host plants, which can withstand attack. Along the way, the scientists discovered intriguing new properties in the forage legume, desmodium. Nutritious for dairy cows, it also repels insect pests of maize and substantially reduces damage from striga, a destructive parasitic weed. In short, the push-pull system improves food security and farm income in an environmentally friendly way – an ideal ingredient in the long-term struggle to reduce hunger and poverty in Africa.[56]

Examples such as these show that sustainable agriculture is not only possible but can be immensely beneficial for both people and environment. Not all ecological initiatives are economically self-supporting, of course. Funded by donations, many NGOs work in partnership with smallholders across the world in a variety of ways, providing education and hands-on participation.[57] Increased educational and research programmes in partnership with the world's smallholder farmers should be a major focus of official development assistance from the developed world.

3.5.3 Developed World Contributions

The industrial farming practices of the developed world can also be improved. In New South Wales, Australia, Gilgai Farms manages 2800 hectares of land,[58] transforming a traditional crop/livestock system into an enterprise of *cell grazing* – a system related to holistic management, originally advanced by Allan Savory. The farm is multifaceted and produces cattle, sheep (primarily for wool), cereal crops, native hardwoods (shelter and timber) and eucalypt mallee plantations (for carbon offsets). Cell grazing mimics the grazing intensity of wild herds by moving stock through multiple paddocks. Critics claim the benefits have not been proven, but actual farms in Australia, the United States, Argentina, Canada and southern Africa have regenerated vegetation in rangeland settings and restored their resilience.

Gilgai Farms is a case in point. They have progressively created perennial pastures, re-establishing native grasses and using organic preparations to foster soil

[56] The International Centre of Insect Physiology and Ecology (icipe) (2015).

[57] For example: Legado; Oxfam; International Institute for Environment and Development, see https://www.iied.org/partnerships-coalitions

[58] Gilgai Farms website. http://www.gilgaifarms.com.au/

microbes and soil life generally. Soil disturbance and costly inputs are minimized and carrying capacity has been increased by about 50%. Whole farm profit reflects this success. The farms' carbon footprint is being minimized, while approximately six tonnes of CO_2 equivalents per hectare per annum are being sequestered.

Harold and Ross Wilkin run an organic farm near Danforth, Illinois, returning excellent profits to investors through Iroquois Valley Farms (IVF), a radical investment company based on long-term capital rather than trading capital. The success of IVF depends on the premium paid by world consumers for organic food, sufficient to balance out the 3 years it takes to convert land to organic status in the United States. It also requires 'patient capital'.[59]

Such ventures can be encouraged by governments in the developed world, such as the Danish government, which announced in 2016 that it intends to double the area dedicated to organic agricultural production by 2020. This initiative, too, is good news for the future of farming.

3.6 Regenerative Urbanization: Ecopolis

3.6.1 Ecopolis: Circular Resource Flows

Section 1.7 described the problems of a steadily growing population and some of the ecological challenges of urbanization. But it also mentioned the important fact that urbanization helps stabilize population.

Following the rationale of co-author Herbie Girardet,[60] cities need to move away from their now essentially linear metabolism, with resources flowing through the urban system without much concern about their origin, or about the destination of wastes. Inputs and outputs are so far treated as largely unconnected. That's one of the downsides of urbanization, which nonetheless has the unintended but highly welcome effect of rapidly reducing birth rates while often making families happier and more prosperous.

Surely, cities need to move towards a circular metabolism, giving plant nutrients – nitrogen, phosphates and potassium – back to farmland, storing carbon in soils and forests, reviving urban agriculture, powering human settlements efficiently by renewable energy, reconnecting cities to the regional hinterland. These measures are the basis for creating viable new urban economies.

The challenge of our time is to transform the highly unsustainable model of today's cities into what Herbie Girardet calls 'Ecopolis', the regenerative city. Cities being our primary home have to comply with the basic laws of ecology (Fig. 3.8).

The 'Ecopolis' model is quite similar to what Agni Vlavianos Arvanitis, also a co-author of this report, has earlier referred to with the 'Biopolis' – an environmen-

[59] Kuipers (2015).

[60] Girardet (2014).

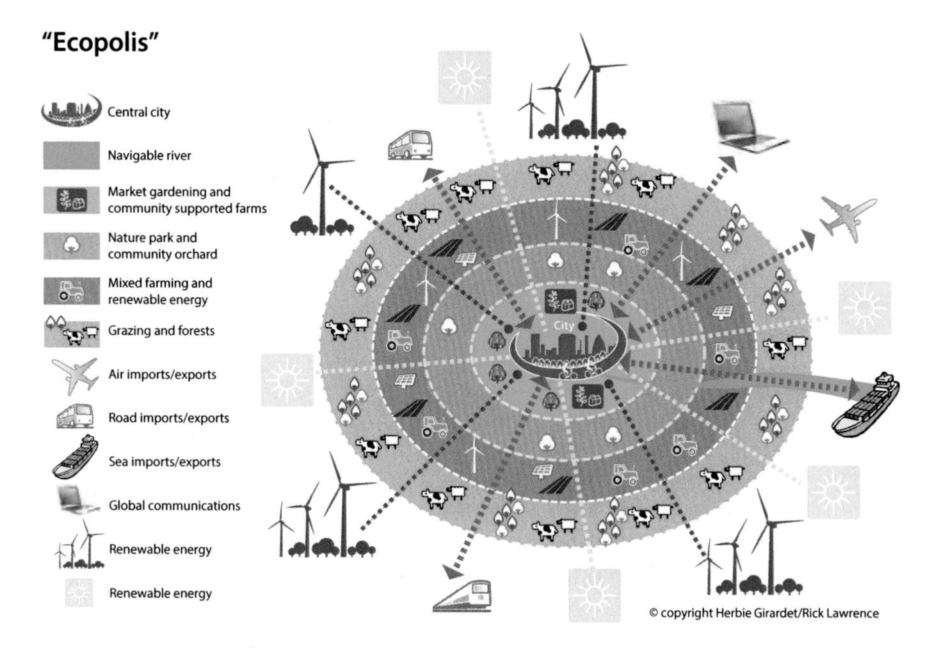

Fig. 3.8 'Ecopolis', the regenerative city, invites many typically rural activities back into the city region, such as market gardening, mixed farming and, most importantly, renewable energies. Fossil fuel dependence and transport intensity is thus greatly reduced (Source: Girardet 2014)

tally sustainable, zero-pollution city, in which human and natural populations live in harmonious balance. It stresses the ethical dimension of a new form of urban life that restores nature and culture to the city: As humans have a collective responsibility, they will be ethically responsible for the damages and problems delegated to future generations.

A recent publication by Germany's WBGU (Scientific Council on Global Environmental Change) delves into these issues in much detail. In its 2016 report on the 'transformative power of cities', written as an input to Habitat III, it states that urbanization does not just pose local sustainability challenges but global ones as well. It is not just a problem that cities often grow on some of a country's most productive farmland, effectively rendering it infertile, but that they make huge demands of a wide range of global resources as well: fuel, food, timber and metals prominently among them.[61]

The issues raised by WBGU were first conceptually addressed by the American urbanist Abel Wolman in his 1965 paper 'The metabolism of cities'.[62] He developed a model which could quantify the inflow and outflow rates of resources of a hypothetical American city of one million people. The benefits of this approach are now widely understood. They enable a clear understanding of urban 'system's

[61] WBGU (2016).

[62] Wolman (1965).

boundaries', which helps explain how cities interact with the natural world. By integrating biophysical and social sciences, Wolman helped to clarify policy and technology options.[63]

This thesis was further developed by Herbie Girardet. He describes the essentially linear metabolism, as indicated above. Nutrients and carbon are removed from farmland as crops are harvested, and then processed and eaten. The resulting sewage, with or without treatment, ends up in rivers and coastal waters downstream from population centres, and the plant nutrients it contains are usually not returned to farmland. Rivers and coastal waters all over the world are polluted by a mix of urban sewage and toxic effluents. Surely, this has to change. Plant nutrients – nitrogen, phosphates and potassium – should come back to the farmland feeding the cities, storing carbon in soils and forests (see Sect. 3.1), reviving urban agriculture, powering human settlements efficiently by renewable energy, reconnecting cities to their regional ecosystems. But with the scale of urbanization now occurring across the world, the prospects for both people and the nature are still barely understood, much less applied.

3.6.2 Regenerative Cities

The concept of the regenerative city is intended to address this question. It is not just about greening the urban environment and protecting nature from physical urban expansion – however important such initiatives are – but about city people taking positive steps to create regenerative urban systems of production, consumption, transportation and construction. Humanity needs to find ways of initiating:

- An environmentally enhancing, restorative relationship between cities and the natural systems they depend on
- The mainstreaming of efficient, renewable energy systems for human settlements across the world
- New lifestyle choices and economic opportunities which will encourage people to participate in this transformation process

A new integrated science of urban planning and management seems to be needed. Conventional urbanistic science, technology and planning are mainly focused on the prospects of gigantic infrastructure investments promising lucrative contracts for firms and glorious careers for municipal rulers. What is missing is an understanding about the relationship between cities and the living world beyond.[64]

In recent years, there have been a great many *urban regeneration* projects in run-down cities of industrialized countries. These have greatly benefitted people immediately affected. But the concept of regenerative *cities* goes further: it focuses on the linkages between city people and nature, between urban systems and ecosystems.

[63] https://en.wikipedia.org/wiki/Abel_Wolman

[64] Batty (2013).

A step into the right direction was taken during the Habitat III (the third United Nations Conference on Housing and Sustainable Development, in Quito, Ecuador, 2016) where the New Urban Agenda (NUA) was adopted. It consists of a wide spectrum of topics on sustainable urban development, compiled in one document adopted by the international community. For the first time, municipalities and cities were officially recognized as key actors of sustainable development.

So far, initiatives towards resource-efficient, regenerative urban development are focussed, above all else, on 'eco-districts' in cities across Europe and the United States. Examples include the Solarsiedlung in Freiburg, Germany; the Beddington Zero Energy Development in Sutton, South London; eco-districts in Nancy, France; Hammarby Sjöstad in Stockholm; and Portland's Eco-District initiative. More often than not, they were made possible by supportive national legislation.

But there are also more ambitious projects that involve the retrofit of entire city regions. Two examples at the end of the chapter offer a few more details.

3.6.3 Cities and Natural Disasters

Another important issue is that cities, as clusters of human populations, can be highly vulnerable to natural disasters such as earthquakes, tsunamis, high tides and floods. Ever greater climate impacts are looming, particularly in river valleys and near sea coasts.

The world's most expensive real estate tends to be located in coastal areas. For instance, of the 25 most densely populated US counties, 23 are in coastal locations. Across the world, major city regions such as New York, Amsterdam, London, Hamburg, Copenhagen, Venice, Tokyo, Shanghai, Kolkata, the Nile River Delta, Dhaka, Bangkok, Jakarta and Manila are all vulnerable to sea level rises of 1 metre, widely forecast by climate scientists by the end of this century. To address this existential problem is, first and foremost, treated as a local issue: huge investments in sea defences and dykes will have to be funded in the coming decades to deal with sea level rises and with more intense rainfall and floods, in particular in huge flat plains where upgrade of dykes will be required for a long distance into inlands from sea coasts along the rivers.

Co-author Yoshitsugu Hayashi[65] stresses the need to avoid building in areas where future disaster risks can be expected, due to climate change. In planning disaster prevention infrastructure, such as sea or river dykes, one should be clear about the height needed to maximize 'resilience', namely, avoidance of catastrophic floods and a deliberate retreat of settlements leaving the high water more room. The 'room for the river' concept adopted in sixteenth-century Japan and in today's Netherlands is an excellent precedent for such measures.

Countries should also take active measures to limit urban sprawl. In planning road and railway infrastructure, they should aim to optimize performance to enhance

[65] Hayashi et al. (2015).

easy access to work places, shops, hospitals or natural parks, while minimizing infrastructure maintenance costs and CO_2 emissions. To implement the downsizing and improvement of transport infrastructure, the efficiency of land use and transport systems across cities needs to be carefully developed.

Future generations must pay more for maintaining the urban infrastructure required while facing ever smaller budgets, most likely due to the ageing population. Our generation ought to plan urban infrastructures avoiding unbearable costs for the generations to come.

3.6.4 Adelaide

At the turn of the century, the people of South Adelaide were getting concerned about dwindling water supplies from the Murray River. South Australia's Labour Premier Mike Rann decided it was time to explore wider sustainability issues in this city region of 1.3 million people. In 2003, Herbie Girardet was invited as a 'thinker in residence' in Adelaide and to trigger discussions of options for combining environmental sustainability and new job creation initiatives.

During hundreds of seminars and lectures over a 10-week period that brought people from all sections of Adelaide society together, many interconnected issues were explored. At the end of his residency, Girardet produced a 32-point plan to help transform the environmental performance of South Australia. During Mike Rann's 8-year premiership, and in the 5 years since, much of this has been implemented:

Forty-five per cent of electricity in South Australia is now provided from wind and solar technology; energy and water efficiency have become mandatory; all organic waste is recycled and returned to urban gardens and farmland on the edge of the city which is also irrigated with recycled wastewater; three million trees have been planted to counter erosion and air pollution; Lochiel Park Solar Village has been built as a model development; thousands of people are working in the city's new, green economy.

In addition to these initiatives, Adelaide has also done much to enhance the liveability of its inner city region. Pedestrian and cycle lanes have transformed the inner city; new tramlines have been built; and much new housing has been made available in converted warehouses and former factories. Adelaide, with the famous Parklands at its centre, is now listed as one of the world's five most liveable cities.[66]

Metropolitan Adelaide, then, has acquired many attributes of a regenerative city. By December 2013, Adelaide achieved 45% of electricity produced by wind turbines and solar PV panels had implemented PV roofs on 150,000 (of 600,000) houses, and on most public buildings, introduced Tindo, the world's first bus running on solar energy, and made solar hot water systems a mandate for new buildings. Furthermore, the city planted three million trees on 2000 hectares for CO_2 absorption as well as biodiversity, and it also built the Lochiel Park Solar Village with its

[66] www.infosperber.ch/data/attachements/Girardet_Report.pdf

106 eco-homes. With all these measures, the Greater Adelaide region has reduced its CO_2 emissions by 20% since 2003. Concerning resource use, Adelaide implemented a zero-waste strategy driven by ambitious recycling incentives. The city collects 180,000 tonnes of compost a year made from urban organic waste and uses reclaimed waste water and that urban compost to cultivate 2000 hectares of land near the city. In this manner, Adelaide has created thousands of new green jobs.[67]

3.6.5 *Copenhagen*

In recent decades, Copenhagen has made remarkable strides towards becoming a liveable as well as a sustainable city, also moving towards being regenerative. The transformation of much of the inner city into a pedestrian zone was the starting point. This has resulted in a 'Mediterranean-style' ambience where markets, cafes and restaurants proliferate. More people cycle in Copenhagen than in most other cities. And initiatives on energy efficiency, combined heat-and-power and renewable energy have gone further than almost anywhere else in the world. The same goes for waste management.

All this started in the 1960s when Copenhagen's City Council decided to establish a huge car-free pedestrian zone around its historical city centre. Pedestrianization was combined with the creation of cycle routes, public transport schemes, combined heat-and-power systems, renewable energy development and recycling projects.

One man who inspired much of this change was Jan Gehl, whose book *Cities for People* has become essential reading for urban design professionals. He is excited by the prospect that by 2025 Copenhagen wants to become the world's first carbon neutral capital city, combining 50 different initiatives that include integrated transport, green architecture, district heating, wind farms in and around the city, electric transport, a smart grid and efficient waste management.[68]

These examples are now being used as models for the regenerative transformation of other cities.

3.7 Climate: Some Good News, But Bigger Challenges

As already stressed in Chap. 1, the world has to undergo a rapid and thorough transformation of its production and consumption systems to have a chance to stay within the 2° target. The Paris Agreement alone and the measures committed by governments so far are far from achieving the goal. Rather than keeping warming below 2°

[67] Girardet, H., Regenerative Adelaide, Solutions Magazine, www.thesolutionsjournal.com/node/1153

[68] Girardet (2014) l.c.

degrees the world is on track for 3° degrees or more. Heating the earth by 2° (Celsius) is not just a little worse than the 1.0–1.3° (or so) we've already warmed it; it's much more dangerous. Three degrees is profoundly more dangerous. Four means living on a frightening, chaotic planet, the like of which humans have never experienced.

So the situation is critical. Yet, let us start with some good news.

3.7.1 Good News

In Sect. 3.4, the exciting trend of a decentralized energy system was outlined, starting with a quote from Amory Lovins: 'Imagine fuel without fear. No climate change…' The chapter goes on showing that renewable energies got ever cheaper over the last 10 or 20 years, meanwhile beating new installations of coal and nuclear. Figure 3.6 showed the seemingly fatal decline of the Dow Jones US Coal index. Investors are switching to renewable energies.

A related development gives additional reason for hope: a broad and worldwide *divestment campaign*, which was mostly motivated by climate concerns. By March 2017, 701 institutions representing $5.46 trillion have sold their shares of fossil fuel companies.[69] It was the fastest growing divestment movement in history.

The accelerating discussion on 'stranded assets' is another sign that change is in the air. As Alex Steffen writes in his blog (March 2017), 'Fuels that can't be burned aren't worth much. In turn, the companies whose major assets are in coal, oil and gas are worth much less than their stock prices would indicate. The difference between the *valuations* of fossil fuel companies and *their true worth* is so large that national banks, financial industry associations and esteemed investors around the world are warning that it represents a bubble potentially as large as the 2007 Subprime Crisis.'

The Barclays Bank, for instance, estimates that limiting emissions to 2 °C would cause a drop in the future revenue of the oil, coal and gas industries of $33 trillion over the next 25 years. The Bank of England published a paper in January 2017 in which it said that the bursting of the 'Carbon Bubble' was 'likely to be abrupt' and 'likely to pose risks to financial stability'.

The price of anything is what you can get someone to pay for it. For investors who own coal, oil and gas companies, supporting the *perception* that these companies will be profitable long into the future is now a multi-trillion dollar priority. This by the way seems to be one of the things that unite Trump and Putin. Both are genuinely interested in keeping the value of fossil assets on as high a level as long as possible.

A different but related matter is CO_2 emissions from transport. But also here, good news are there, reported by the Carbon Tracker Initiative and Grantham Institute at Imperial College London. In essence, the scenarios presented assume a

[69] https://gofossilfree.org/commitments, accessed March 13, 2017.

steep rise of solar (PV) electricity and, in parallel, of electric vehicles.[70] If this would happen, it is likely to halt growth in global demand for oil already from 2020 and onwards and lead to the stranding of fossil fuel assets as the low-carbon transition gathers pace. The result could be more or less carbon-free mobility a few decades from now. But it hinges, of course, upon the phasing out of coal as a major fuel for electric power generation.

In Sect. 3.9, evidence will be added of the huge potential of energy savings. A fivefold increase in energy efficiency seems possible and would dramatically reduce the need for energy supplies. However, to establish commercial profitability, some major changes of frame conditions will be required, as is discussed in Sect. 3.10.

Completely different good news come from another corner. It was chiefly the 9-year-old German boy, Felix Finkbeiner, who in 2007 began thinking big about tree planting. Learning about the threats of global warming and hearing about Wangari Maathai and the Green Belt Movement planting 30 million trees in Kenya, Felix reasoned that the children of the world could join and plant many, many more trees. In the same year, the Plant-for-the-Planet initiative was founded, beginning with a commitment of planting one million trees in each country of the world.

The movement grew faster than ever expected. They organized 'academies' for children aged 8–14, empowering them to become Climate Justice Ambassadors. By 2016, roundabout 51,000 children from 193 countries received that title. The goal for the movement today is that every world citizen, on average, should plant 150 trees to reach 1000 billion trees until 2020. That would help absorb a significant part of CO_2 emissions.

One further encouragement for climate action is the nexus with agriculture, touched upon in Sect. 3.5. Restoring soil to high fertility is obviously helpful for high agricultural yields. But it also dramatically increases the capacity of soils to absorb CO_2 (see also Sect. 3.1.4). This means that the task of feeding a world of 7.5 billion people needs not conflict with climate policy goals, with the caveat that the number of cattle should rather be reduced than increased, due to the methane emissions resulting from their digesting.

3.7.2 Addressing the Historic Debt and the 'Carbon Budget' Approach

The Paris treaty is a call for action to all governments in the world. However, the necessary changes must start in the industrialized countries. They built their standard of living on cheap oil and gas and owe it to the developing world to lead the way.

Industrialized countries are, of course, just part of the puzzle. Whether the Paris targets will be met or not will to a large extent be decided by trends in the developing

[70] Carbon tracker Initiative (2017).

world. But developing countries depend a lot on the use of technology currently available mostly in the industrialized countries (China and a few other developing countries being the exception). They also need to see good examples that welfare and well-being can be achieved in a low-carbon economy.

The North-South talks in the climate negotiations often circle around money transfers from the North[71] to low-income countries in the South. The Paris commitment that has been $100 billion yearly from 2020 is also to be used for adaptation to an increasingly changing climate. This sum is still modest compared with the global subsidies to fossil fuels which are about five to six times higher. The practical problem is, however, the perception by most governments and parliaments in the North that they have almost no room of manoeuvre in their public budgets. The real wealth of those countries tends to be in private hands.

This fact can lead to a different strategy for making the transformation towards a low-carbon economy possible. A compelling idea for such strategy was developed already in the 1990s by the late Anil Agarwal and his colleague Sunita Narain[72] from India: The authors proposed to allow everybody on earth the same quantity of resource consumption or of emitting greenhouse gases into the atmosphere. Poor people could sell some of their allowances to rich ones, alleviating their poverty and yet maintaining a strong incentive for both rich and poor people to become more resource efficient and to reduce their carbon footprints. Regretfully this 'one person – one equal allowance' idea never obtained the necessary support.

More than a decade later – and with a view of facilitating climate negotiations at COP 15 in Copenhagen – the German Advisory Council on Global Change ('WBGU') further developed the idea and introduced the 'budget approach',[73] schematically explained in Fig. 3.9. This approach meant to give countries of all types the same 'budget' of carbon emissions per capita. The old industrialized would be forced to go shopping for permits in the less developed countries.

The exciting feature about this budget approach is the following: for the first time in history, a developing country standing before the decision of building a new fossil power plant would not automatically go for it but would lean back for a moment and then calculate the cost-benefit relations for two options – building or not building. High prices for carbon permits would make the non-building option temptingly lucrative. And if plenty of options still exist of expanding renewable energies (Sect. 3.4) or energy efficiency (Sect. 3.8), the balance would swiftly turn to the non-building option. And that for purely economic reasons.

Unfortunately for the climate negotiations, the United States, Russia, Saudi Arabia and a few others came to the Copenhagen climate summit with the clear intention to block the discussion on the budget approach. For the Club of Rome, however, it looks very attractive and worth reviving.

---

[71] By "North" we include Australia and New Zealand.

[72] Agarwal and Narain (1991).

[73] WBGU (2009).

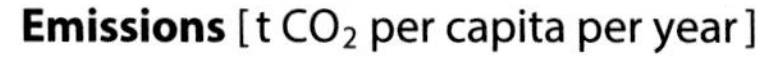

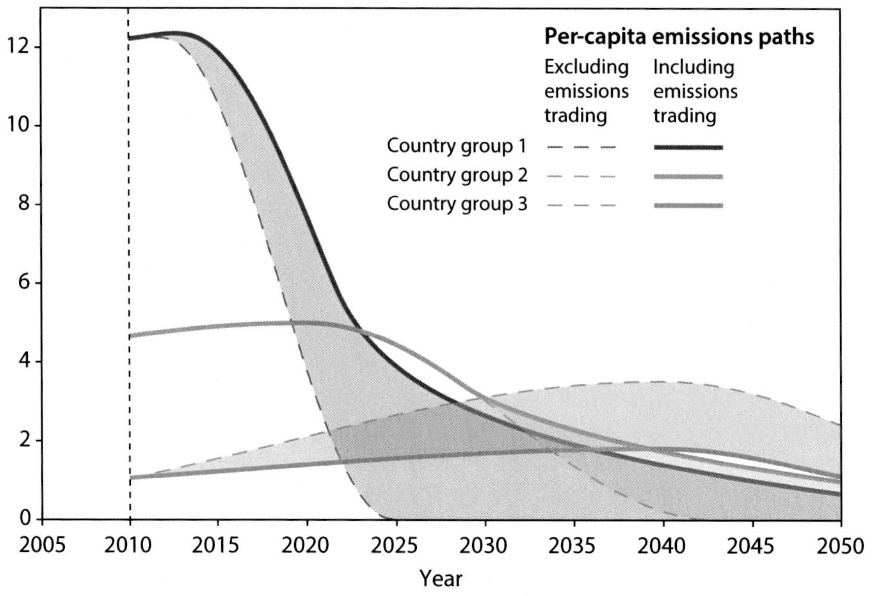

Fig. 3.9 The 'budget approach': rich countries (*pink*) have almost exhausted their budgets of CO_2 emissions. The dotted lines show budget developments before trading sets in. The developing countries (*green*) would have an excess of permits and could sell some – allowing the rich countries still to emit CO_2. The middle income countries (*yellow*) can also buy permits after their budget may have shrunk to zero in 2040 (Source: WBGU – German Advisory Council or Global Change (2009): Solving the climate dilemma: The budget approach. Special Report. Berlin: WBGU)

3.7.3 A Price on Carbon

The budget approach is a means for international deals. On the domestic scales, trading permits are a lot less attractive, as the experience has shown with the European Union (EU) ETS (Emissions Trading System). The price of emission permits has been – and still is – far too low to make a difference. In practical terms, carbon taxes are much easier to handle and more effective. The trouble is that politically, they tend to be seen as 'toxic', notably in the United States. One attractive way forward would be to follow Jim Hansen's suggestion, and more recently also proposed by the new (Republican) Climate Leadership Council (CLC), to impose a carbon tax but refund the money to the taxpayers, on an equal and quarterly basis via dividend checks, direct deposits or contributions to their individual retirement accounts.[74] If such action would be taken the incentives to invest in alternative energy and fossil-free industrial processes would get a further strong boost.

[74] Thorndike (2017).

One problem with all taxes and carbon trading systems is that they either badly hurt carbon emitters (politically extremely difficult) or are so tame that they don't really help the decarbonization of the economy. One proposal trying to combine the advantages (politically acceptable and yet with a mighty steering effect) is discussed in Sect. 3.12.3: a gradual increase of the price in proportion with documented efficiency increases, so that annual expenses for carbon or energy services remain stable, on average.

3.7.4 Combatting Global Warming with a 'Post-war Economy'

Clearly, the practical steps done so far by governments and private actors are far from sufficient to meet the Paris goals. In response, an increasing number of commentators, including climate scientists, argue in favour of a massive, *war-like mobilization* to win the battle against climate change. Hugh Rockoff, a Rutgers University economics professor, draws parallels between the fight against climate change and WWII.[75] According to Rockoff, the scale of our financial hurdle in fighting climate change is similar to that faced by our parents and grandparents during WWII. The way in which they accomplished this – and what Rockoff suggests we will have to do to beat global warming – would be to pursue tremendous government spending on infrastructure and technology.

The implication is that the time of 'incrementalism' is over. What we need now is transformation through technology innovation, substitution and large-scale investments. Here governments must play a key role.

As Club of Rome, we would rather avoid the term 'war-like mobilization', so let us instead use the term 'post-war economy'. The United States, and also the countries defeated in WWII, Japan and Germany, experienced a massive economic upswing after the war by building (or rebuilding) infrastructures and developing new technologies.

While politically working on changing the frame conditions conducive to drastic changes – like moving into a 'post-war economy' and/or adopting the budget approach – it remains necessary to pursue sector-based options, some of them exciting, like renewables, efficiency subsidies, smart mobility, farm reform, slowing down forest cutting, etc. Policy frameworks have to be changed to incentivize the necessary technology shifts. In addition, public sector support for research, innovation and demonstration projects has to increase significantly. Moreover, public procurement – in many countries representing one-fifth of GDP – should be used proactively to promote low-carbon solutions. Of crucial importance will be to support investments in low-carbon infrastructure and material efficiency. Furthermore, it will be necessary to oblige the financial industry to report on the carbon risks of their lending.

[75] Rockoff is quoted by Tartar (2016).

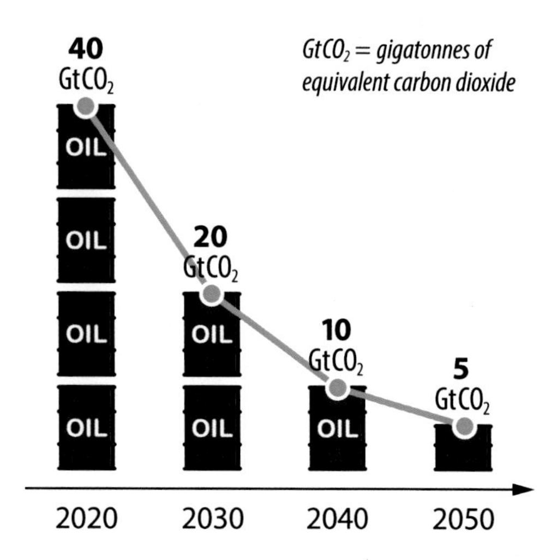

Fig. 3.10 Roadmap of massive emissions reductions as suggested by Johan Rockström et al. (2017)

Innovation has to give much greater priority to public goods, in this case low-carbon solutions. In our view, innovation activities today are too much dominated by greed and quickest possible return on investment. Governments ought to significantly increase their funding of research and innovation for low-carbon solutions. But under the conditions outlined in Sect. 3.10, of steadily and predictably rising carbon emission or more generally energy prices – preferably using a carbon tax – both governments and private investors would almost automatically change their priorities exactly in the desired direction.

A few of the world's most renowned and respected climate experts – among them Johan Rockström and John Schellnhuber – challenged conventional wisdom in an article.[76] The authors state that 'although the Paris Agreement's goals are aligned with science and can, in principle, be technically and economically achieved, alarming inconsistencies remain between science-based targets and national commitments'. They are afraid that long-term goals will be trumped by political short-termism. Hence, they put forward a roadmap in the form of a 'carbon law' – seemingly inspired by Moore's law – that would imply a halving of carbon emissions every decade until 2050. By following such a pathway, GHG emissions would be close to zero in 2050, a precondition for reaching the 2° target with high probability (Fig. 3.10).

The roadmap affects all sectors and suggests much more rapid action than hitherto discussed. Fossil fuel subsidies have to be abolished no later than 2020. Coal has to exit the energy mix no later than 2030. A carbon levy at minimum 50 US$/

[76] Rockström et al. (2017).

ton must be imposed. Combustion engines should no longer be sold after 2030. After 2030 all building construction must be carbon-neutral or carbon-negative. Agro-industries must develop sustainable food strategies and massive reforestation programmes be launched. Removal of CO_2 from the atmosphere will have to complement in the form of BECSS and/or direct air CCS (DACCS).

The main focus of climate mitigation so far has been on energy use. But material throughput in society is equally important. Recent studies on global material stocks and flow project[77] up to a fourfold increase in material stocks by 2050. To limit the carbon footprint, Krausmann et al. argue, will require 'rigorous decoupling of services from material throughputs' (see Sects. 3.8 and 3.9).

The Paris Agreement does not address emissions associated with land-use changes other than forestry. Soils are the largest natural terrestrial store of carbon. Yet, so far agriculture is not part of the mitigation agenda. Agriculture represents a daunting challenge. Every calorie of food on the table today is backed up by at least five calories of oil. Less than 100 years ago, the relation was the opposite: each calorie of technical energy going into farms yielded roughly five calories of food – owing to the generous energy input from the sun. Here somewhat of a revolution must take place, by replacing fossil fuels with smart biofuels, reducing the amounts of fertilizer and pesticide used and by building carbon in the soils. Low-till farming, perennial crops, cover cropping and crop rotation are activities that increase soil organic matter. As said before, the availability of this transition is extremely good news.

The loss of forests contributes between 12% and 17% of annual GHG emissions. Replanting efforts have accelerated in recent years and deforestation has slowed down. But much could still be done, inter alia, using the admirable youth energy of the *Plant for the Planet* movement. Also some new technology could be applied – as suggested by Club of Rome member Agni Arvantitis – like using drones, firing small pods that contain germinated seeds into the soil.

Biofuels can be part of the solution. In forested countries like Canada, Sweden and Finland, the concept of a bio-based economy is gaining ground. Petroleum-derived products will be replaced by better-performing products based on renewable materials from forests and agriculture. Bio-energy is but one of many product streams. Of particular interest would be to explore the potential of algae. They can be grown almost everywhere and do not require cropland. Algae use nutrients more efficiently and can produce over ten times more oil per acre than typical fuel plants.

All those measures should be accompanied by a new civilizational mind-set of sufficiency. The extremely high political priority for jobs is always in danger of artificially creating 'hamster wheel' activities with pretty little satisfaction to those treading the wheel and also to those consuming the often silly products from the 'wheel'. To achieve the Paris goals, consumers must assume a much more important role. To foster a new consumption culture, however, one prerequisite would be the development of new indicators. To replace growth of GDP with quality indicators is one type of measure. Another will be to develop indicators for the true carbon footprint of each individual.

[77] Fridolin Krausmann et al. 2017. Global socioeconomic material stocks rise 23-fold over the twentieth century and require half of annual resource use. 114 no. 8.

Today's emissions statistics rely on 'production-based' accounting. But territorial emissions are only part of the picture. As an illustration, let us look at Sweden. Per capita emissions based on production-based accounting are less than six tonnes. Consumption-based accounting, including international air travel, on the other hand, puts per capita emissions at ten tonnes. In this regard, tracking carbon emissions through supply and retail chains and insert labelling at the point of sale or use would be a first necessary step.

3.8 Circular Economy Requires a New Economic Logic

Today's economy builds on a 'fast turnover' principle – take, make and dispose. The faster we replace our gadgets, the better it seems; and this now applies to most of the items we consume – from cheap clothing to expensive cell phones. The construction sector – accounting for 30–40% of material throughput in society – is no exception. Here, as well as with consumer goods, the way we manage earth's resources is both grossly inefficient and the generator of large volumes of waste. The consequences are rapidly increasing levels of pollution, resource depletion, loss of vital ecosystems and substantial losses of economic value with each product disposed. To continue on this path would lead to a disastrous situation over time. The combination of resource depletion, a changing climate and pollution would restrain growth and ultimately bring the economy down.

The economic values lost because of the linear material flows are seldom talked about. In Europe, for instance, most of the value of input materials is being lost after one use cycle, in spite of valiant recycling efforts ('Growth within', McKinsey 2015). Even in the best systems, far from all materials produced are reused or recycled, and those that are recycled very often cannot be used again because of poor design, contamination or the lacking of standards. Electronics is a pertinent example. The design is such that most electronic products cannot be dismantled. Another example is the high-quality steel in cars. It becomes so contaminated in the scrapping process that it is primarily used as low-value construction steel. The same story can be said about many other materials – plastics are a case in point. The implication is that most secondary materials are either incinerated, landfilled or only used in low-value applications.

Also from a strict climate change mitigation perspective, the present linear economic model is highly problematic. We know that recycling and reuse of materials, not least metals, save a lot of energy – and hence pollution. The extraction and production of basic materials – like steel, cement and aluminium – account for almost 20% of global GHG emissions. Switching to renewables and improved energy efficiency in the production processes would help. However, it is just as important to reduce material throughput through activities like reuse, recycling, extended product life, remanufacturing and through innovation and product substitution. Given that demand for basic materials is expected to increase rapidly in the future – half of the urban infrastructure that will be needed in 2050 has not been built yet – a revolution in the way we use basic materials and their substitutes is urgently needed.

3.8.1 The Economy Must Be Transformed

Natural resources provide the basis of prosperity and well-being. All the UN's Sustainable Development Goals (SDGs) depend on the sustainable management and use of earth's natural resources. The relationship was made abundantly clear in a landmark report – *Resource Efficiency: Potential and Economic Implications* – by the International Resource Panel (IRP) , launched at the G7 meeting in May 2016 in Japan.

The IRP report explains in detail the risks faced with today's more or less linear production systems. It advocates a radical shift in mind-sets as well as production and consumption systems. Unless resources of all kinds are used much more efficiently, the SDGs *will not* be met.

But enhanced resource efficiency is but a step in the right direction. Equally important will be to move towards an economy based on renewable materials, circular material flows and where taxes are used to balance demand. If not, the efficiency gains made will be rapidly eaten up by a combination of rebound effects and economic growth. Regretful as it is, most policy interventions in the past overlooked these aspects and thus failed to bring about absolute decoupling.

Governments and businesses must work together to develop resource strategies to avoid both resource constraints and increasingly serious problems of waste and pollution. The productivity concept must be broadened to include natural resources. Labour productivity increased by at least a factor of 20 since the birth of the Industrial Revolution, while the increase in resource productivity has been modest. Since the year 2000, it has in fact decreased, if looked at from a global perspective. What we experience today is 'resource recoupling' rather than 'decoupling', that is, the growth in resource demand is larger than the rate of economic growth. At a time where labour is highly productive and unemployment has become a worldwide scourge, shifting efforts to the productivity of basic resources, like energy, materials, soil and water would make a lot of sense.

A new business logic is needed. Circular business models must replace linear ones. One particular challenge for the future would be to generate a breakthrough for the concept of *services instead of products* for a wide array of consumer products – such as computers, cell phones, household appliances, cars, furniture and textiles. Even in the property market, the same principles could apply.

A crucial issue will be: How can the principle of 'earning revenue by selling more stuff' be replaced by a system where revenue increasingly is the result of quality of service on products that last?

One of the pioneers of the concept, Club of Rome member Walter Stahel, puts it the following way: 'Societal wealth and well-being should be measured in stock instead of flow, in capital instead of sales. Growth then corresponds to a rise in the quality and quantity of all stocks – natural, cultural, human and manufactured. For example, sustainable forestry management augments natural capital, deforestation destroys it; recovering phosphorus or metals from waste streams maintains natural capital, but dumping it increases pollution; retrofitting buildings reduces energy consumption and increases the quality of built stock'. From this reasoning follows yet another argument for replacing growth in GDP with indicators giving prominence to quality, not quantity.

The transition to a new business logic will require decisive policy action. The cost structure of the economy is seriously flawed. Financial capital is overvalued while both social capital and natural capital are undervalued. Unless these flaws are acted upon the circular economy will not materialize.

Fortunately, calls for a new model of production and consumption are becoming increasingly frequent, spurred by a series of studies by the Ellen MacArthur Foundation, the EU Commission, OECD, the World Economic Forum and the Club of Rome. In the EU, the legislative proposal 'The Circular Economy Package' was presented in December 2015 and is now being examined and debated by Member States' governments and the European Parliament.

Studies by the Ellen Mac Arthur Foundation, the EU Commission and the Club of Rome have highlighted the fact that moving towards a circular economy – by using and reusing materials, rather than using up – would yield multiple benefits. The proposition is that a circular economy, where products are designed for ease of recycling, reuse, disassembly and remanufacturing – and where products and property are used much more efficiently, for example, through leasing and sharing – should replace the traditional, linear 'take, make and dispose' model that has dominated the economy thus far.

An economy favouring reuse and recycling of materials as well as product-life extension is, by definition, more labour intensive than one based on a *disposal* philosophy, that is, linear resource flows. Caring for what has already been produced will result in more jobs being created than both mining and manufacturing in often highly automated and robotized facilities.

3.8.2 The Societal Benefits of Moving Towards a Circular Economy

A Swedish case study from 2015[78] illustrates that moving towards a circular economy would contribute significantly to building economic competitiveness, increasing jobs and cutting carbon emissions. Subsequent reports, covering seven more European countries (Finland, France, the Netherlands, Norway, Poland, Spain and the Czech Republic), look at the effects of three decoupling strategies that underpin a circular economy – *increasing the share of renewable energy and enhancing energy as well as material efficiency.* The studies use a traditional input/output simulation model and conclude that by 2030, carbon emissions could be cut by between 60% and 70% in all of the countries examined if a key set of policy measures were implemented. The effects on employment vary between the countries studied, but the number of additional jobs is in the range of 1–3% of the labour force.

[78] The Club of Rome (2015).

The report considers a number of policy options and investments that would help advance a circular economy and the climate and job benefits that it would bring about:

- Address the flawed cost structure of the economy by letting market prices bear their full costs.
- Rethink taxation – in favour of a tax shift, lowering taxes on labour and increasing taxes on the use of nature. (Such a tax shift would accelerate the transition to a circular economy. It would also help balance the threat of losing jobs in an increasingly digitized economy.)
- Strengthen recycling and reuse targets to help reduce and process waste and residues. Put limits to waste incineration.
- Strengthen existing policies of promoting renewable energy, such as feed-in tariffs and green certificates.
- Introduce design requirements for new products for ease of repair and maintenance, dismantling and countering obsolescence. Introduce, as well, material and product standards in key sectors of the economy.
- Use public procurement to incentivize new business models, moving from selling products to selling performance.
- Make material efficiency a core part of climate mitigation policies. Most climate change mitigation strategies are sector-based, with a primary focus on energy use. But the Club of Rome study referred to demonstrates the benefits in terms of significantly lower carbon emissions from using products longer and from enhanced rates of recycling and reuse.
- Launch investments, primarily in infrastructure, to support the circular economy.
- Support innovation in low-carbon solutions.
- Exempt all secondary materials from VAT.

Developments in the EU are crucial. No nation can close material loops by itself. Common rules at the EU level, however, would move the agenda forward significantly. A problem so far (June 2017) is that the EU Commission, while launching the CE package, has refrained from any meaningful action with regard to the very issues that will decide whether a move towards a more circular economy will happen, that is, a shift in the tax base, the provision of design requirements for new products as well as product standards. Most efforts hitherto have been devoted to changes in the waste directives. Increased levels of recycling rates will have limited effect, however, as long as the great majority of products put on the market are not designed for effective reuse and recycling. When products are difficult to dismantle or there are too many different material qualities – like is the case with plastics and most construction materials – the secondary materials market will not function well. The result will be that most recycled materials will end up as waste or in low-quality applications.

What is urgently needed are policy measures that will incentivize companies to put products on the market that are designed to be recycled or reused at the end of their useful life. In the EU context, the eco-design directive – which so far has mainly aimed at promoting energy efficiency – would be well placed to enhance material efficiency as well. Taxation ought to be considered as a policy instrument as well. Current taxation laws do not reward companies that take a circular economy approach. VAT rates could easily be based on life-cycle analysis of the environmental impact of products or favour products with a high content of recycled materials. Last but not the least, a tax shift – lowering taxes on labour and increasing taxes on the use of nature – would significantly help bring about the transformation of the economy that is urgently needed.

3.9 Fivefold Resource Productivity[79]

Enhancing resource productivity was mentioned as a very reasonable thing to do for reducing unemployment and for decoupling well-being from resource use. In *Factor Five*, a 2009 report to the Club of Rome, the authors demonstrated that a fivefold increase in resource productivity is available *even* in the four most energy- and water-intensive sectors (construction, industry, transport and agriculture).[80] The book also pointed out that much of this potential remains dormant, largely due to low resource prices. It is very encouraging, however, that even under mostly unfavourable conditions, considerable progress is being made and can be observed.

3.9.1 Transport

Transport is a difficult and key sector with regard to carbon productivity. In *Factor Five*, three main areas were presented as means of achieving significant reductions in greenhouse gas emissions, namely, *changing to low/no carbon energy sources for vehicles, increasing the energy efficiency of vehicles and ensuring appropriate modal choices*, such as mass transit rather than individual passenger vehicle commuting.

Oil-based liquid fuels are unlikely to play a long-term role in mobility. As a consequence, engineers have worked hard on cost-effective alternatives, with efforts ranging from alternate liquid fuels that capitalize on existing distribution infrastructure, to replacing internal combustion engines with electric motors. In 2012, Tesla Motors released their Model S and became the global leader in electric vehicles practically overnight. Since, nearly all major manufacturers are offering electric cars on the market. Obviously, in terms of CO_2 emissions it is useless turning to electric drives if that power comes from coal burning utilities. So a precondition for electrification of the vehicle fleet will be for power generation to become low-carbon.

Many more advances are needed in both the automotive technology and the infrastructures supporting resource-efficient transport modes.

---

[79] This Chapter was drafted by Karlson 'Charlie' Hargroves, Daniel Conley, Nestor Sequera, Joshua Wood, Kiri Gibbins, and Georgia Grant, based at the Australian Curtin University Sustainability Policy Institute (CUSP) and the University of Adelaide Entrepreneurship, Commercialisation, and Innovation Centre (ECIC).

[80] von Weizsäcker et al. (2009).

The *Factor Five* examples emphasize that it is never one single fix but rather the whole system design that raises the potential for efficiency gains. Studies show that a reduction in the weight of a vehicle by as little as 10% can improve fuel economy by 6–8%.[81] One of the easiest ways to achieve this is by using alternatives to steel to reduce the weight of the vehicle where appropriate.

According to the US Energy Information Administration, weight reduction and aerodynamic advances can deliver a 45% fuel demand reduction for heavy vehicles and are expected to deliver a further 30% by 2030 through complimentary technology improvements.[82]

More gains can be expected from appropriate modes of mobility. In effect, this means reducing automobile dependence.[83] Modal shifts can be achieved by influencing the economics of mobility in favour of a preferred mode of transport such as rail. One method of inducing large numbers of commuters to use rail and buses is congestion charging, a toll for driving in certain parts of a city, which applies to particular motorways or over whole zones. London introduced it in 2013 and reduced congestion by 30% in the first 12 months while reducing emissions by 16%. Some £1.2bn of the net revenue generated by the scheme has been invested directly into public transport, as well as into walking and cycling infrastructure.

Together with efforts to discourage vehicle use, many cities are now investing heavily in rail infrastructure, both in light rail for passengers and heavy rail for high speed and freight. As of 2012, the construction of rail systems was underway in 82 cities in China, and in 2016 the China Railway Corporation announced plans for rail projects in another 45 cities. In 2015, plans for rail systems in 50 Indian cities were confirmed. Fast electric rail services cost around the same per kilometre as most freeways, and while they are most effective in densely populated zones, they can also be implemented into wider-spread car-dependant suburbs.[84] An example of this is Perth's Southern Rail line which since opening in December 2007 is carrying 80,000 passengers per day compared to the previous bus system which handled just 14,000.

There is particular potential to achieve a greater than factor five transformation in the freight industry, which in the United States accounts for around 9% of the nation's greenhouse gas emissions. Shifting long-haul freight movements from trucks to rail has the potential to reduce freight-related transport emissions by 85%, a figure which also accounts for the truck transport required at either end of the journey.[85]

The International Energy Agency's (IEA) *avoid and shift* policies for travel include a combination of land-use considerations, transport planning options and modal shifts. The recommendations include case-specific options for particular cities due to different features, such as bus rapid transit, urban cycling, transit-oriented development, mobility and transport demand management, carpool incentive programmes, tele-work programmes, parking policies and shifting long-distance pas-

---

[81] Pyper (2012).

[82] RMI (2011).

[83] Newman and Kenworthy (1989).

[84] Newman and Kenworthy (2015).

[85] Frey and Kuo (2007).

senger travel and freight to rail. This scenario is estimated to deliver some US$ 20 trillion in global savings through mitigated infrastructure spending by 2050,[86] with the potential to mitigate up to 50% of global urban transport emissions.[87]

3.9.2 Resource-Efficient Buildings

Buildings and the associated energy used for the production of their electricity and for heating accounted for over 18% of global greenhouse gas emissions in 2010. The best results in reducing emissions are achieved through a focus on space heating and cooling, domestic hot water, appliances, lighting and refrigeration. The main residential case study is the concept of 'Passivhaus', a German innovation from the 1990s. It is essentially heated passively by solar radiation and the heat produced by occupants and appliances, and fulfils the following minimum performance criteria:

- Annual heat and cooling requirements of less than 15 $kWh/m^2/year$.
- Very low building envelope air gaps (tested by a blower door test).
- Primary energy consumption is less than 120 $kWh/m^2/year$.

The 'Passivhaus' concept draws on advanced insulation and airtightness in combination with heat exchange ventilation to deliver fresh air all year round with minimum heating energy requirements. One example is the German Heidelberg-Bahnstadt development, which includes over 1000 apartments designed to Passivhaus standards and is serviced by a district heating system – achieving an 80% reduction in heating energy needs. This concept is gaining traction around the world, with the United States now boasting certified houses, schools and commercial retrofits. The Centre for Energy Efficient Design in Franklin County, Virginia was the first public school (K-12) in the United States designed to Passivhaus standards. It also uses on-site renewable energy generation making it carbon-negative, which means it produces significantly more energy than it consumes.

In recent years, 'Green Buildings' have entered the mainstream with many commercial structures achieving significant reductions in the energy and water consumption. As of 2014, there have been more than 700 'Energy Star' rated commercial building projects providing estimated cost savings of US$ 75 million that also means GHG emission reductions of 600,000 $MtCO_2$. An Australian study indicated that improving energy efficiency through simple measures can result in energy savings of at least 50%, an amount that can save AUD 10,000 (ca € 6800) annually for an average 2500 m^2 office space.

One example is the Pixel Building (Fig. 3.11) in Melbourne, Australia, that does not produce carbon emissions because of its innovative energy use. The building is designed to have 100% water self-sufficiency, implements a no recirculation air system and utilized a new mix of concrete called 'Pixelcrete', which approximately halves the embodied carbon in regular mix. Sixty per cent of the cement is replaced

[86] IEA (2013).

[87] Creutzig (2015).

Fig. 3.11 Pixel Building, Melbourne. Grocon's Pixel building is the first carbon neutral office building of its type in Australia (Courtesy studio505, Dylan Brady & Dirk Zimmermann. Melbourne, Australia / photo: John Gollings)

with pulverized blast furnace slag and fly ash, as well as 100% recycled and reclaimed aggregate. In addition, the building will offset the embodied carbon emissions generated during its construction over its 50-year life cycle through surplus renewable energy produced on site and fed back into the electricity grid.

Concrete is a key product with high energy intensity, which in Australia alone represents more than 20% of residential and as much as 63% of commercial building related embodied energy. Combined with the systematic use of recycled concrete the switch in cement type can deliver fivefold reductions in energy per kilogram.

So for example, construction projects around the world are now using geo-polymer concrete, the largest being Brisbane's West Wellcamp Airport (BWWA) which has approximately 25,000 m^3 of aircraft pavement grade concrete, and 15,000 m^3 geo-polymer concrete used elsewhere on site (total of 40,000 m^3 or 100,000 tonnes). Using a geo-polymer concrete saved the project 8640 tonnes of CO_2 emissions.

3.9.3 Water Efficiency for the Farm

Agriculture was responsible for over two-thirds of the world's freshwater consumption and 14% of global greenhouse gas emissions in 2010, with both of these figures set to rise due to a continual growth in demand for food. Regulated Deficit Drip Irrigation (RDDI) and Partial Root-zone Drying (PRD) fare potential areas of improvements in water productivity in agriculture, which stand to save irrigation water by up to 50% with minimal or no impact on crop yield. Since 2010, 'no-till cropping' has also emerged; it promises to achieve further increases in water and energy efficiency for farms.

A Regulated Deficit Drip Irrigation (RDDI) strategy controls irrigation pattern. It induces increased yields by holding back on water when growth is slow and giving ample watering during times of rapid growth. For example, in the cool temperate environment of Tasmania, Australia, RDDI has shown the potential for a 60–80% reduction in water use on dairy pastures. This could increase the average pasture response to irrigation across the industry by up to 90%. Grapevine growers in the South Australian wine region have achieved a 90 and 86% increases in water use efficiency growing Riesling and Shiraz grapes, respectively, using an RDDI approach.

Audits of irrigated farms have found that energy used in irrigation can account for upwards of 50% of a total farm energy bill. The use of irrigation management systems like CIMIS (the California Irrigation Management Information System) gives real-time irrigation advice to farmers to minimize the overuse of water to irrigate crops.

Similarly, through the use of the online weather system technologies which provide temperature, rainfall, moisture, dew and solar radiation data, tomato farmers in Brazil have managed to halve their water use (from 800 mm/ha to 400 mm/ha) by making irrigation and chemical applications more efficient. This concept combined with an efficient irrigation system can yield a 60–70% reduction in the cost of energy used to pump water. However, uptake from farmers has remained slow and the array of benefits from these strategies are still waiting to be achieved on a large scale.

3.10 Healthy Disruption

The preceding chapters are optimistic. But notably with regard to climate, a lot more powerful action is needed than has been seen in the recent past. In a sense, our economic and social systems require nothing less than disruptive improvement. In Sect. 1.11, some problematic, even frightening sides of disruption, digitization and exponential developments were discussed. Such downsides must be kept in mind when the positive sides of disruption are addressed. It may help to term what is needed *healthy disruption*.

3.10.1 Thirty Years of Welcoming IT

Section 1.11 showed that the digital revolution started in sync with the popularization by the Brundtland report of the concept of sustainable development (SD). The Brundtland Commission anticipated the potential of ICT developments and expected them to make a significant contribution to sustainability.[88] The International

[88] June 2012. "ICTs, the Internet and Sustainability: An Interview with Jim MacNeill". IISD.

Telecommunications Union (ITU), a UN agency, organized the World Summit on the Information Society (WSIS, 2003/2005) which claimed 'that the ICT revolution can have a tremendous positive impact as an instrument of sustainable development'.[89] The WSIS's 'Declaration of Principles'[90] repeatedly mentioned SD, called international agencies to 'develop strategies for the use of ICTs for SD, including sustainable production and consumption patterns' and listing fields of activity in which ICT applications could facilitate SD.

Thirty years after the Brundtland report, digital technologies have expanded at a very rapid rate, fuelled by a combination of increasing computing capacities of microprocessors, positive externalities of telecommunication networks and low marginal costs of data expansion and of adding new nodes to the networks. Improving electronic services is also inexpensive once the backbone infrastructures are in place. These effects have allowed for extremely fast ('disruptive') expansion of IT services at rapidly falling cost and still made the innovators and patent owners such as Mark Zuckerberg billionaires in an extremely short time (compared to the accumulation of wealth at the time of the Rockefellers). But the potential of this disruption for sustainable development waits to be fully realized.

3.10.2 'A Good Disruption'

A very balanced and mostly optimistic view of digitization and more specifically of disruptive technologies and their usefulness for sustainable development comes from Martin Stuchtey et al.[91] In particular, their book emphasizes the usefulness of big data for the energy transition and the circular economy, notably for the recovery of valuable resources that are otherwise lost in the waste stream.

The authors give three major examples for digitizing the physical world that echo the topics covered in Sect. 3.9, and that in a sudden, disruptive manner: mobility, nutrition and housing. For mobility, they recount the story of Uber and other electronic transport services signalling the era of sharing instead of ownership, the electrification, autonomous driving, and lightweight technologies that reduce the ecological footprints of vehicles. A mere 10 years ago, almost no one would have imagined all this disruption. In terms of food production, new techniques include precision agriculture,[92] closing the nutrient loops, and the restoration of natural capital, including the famous restoration of China's Loess plateau of 1.5 million hectares – lifting 2.5 million people out of poverty. Again hardly imagined by anybody a mere 10 years ago. In housing, we now witness spectacular giant 3D printers

[89] "Tunis Commitment". paragraph 13. http://www.itu.int/wsis/docs2/tunis/off/7.html

[90] "Geneva Declaration of Principles". http://www.itu.int/wsis/docs/geneva/official/dop.html

[91] Stuchtey et al. (2016).

[92] e.g. https://soilsmatter.wordpress.com/2015/02/27/what-is-precision-agriculture-and-why-is-it-important/

in Suzhou, China, capable of building houses in 24 h, at an estimated cost of $5000 each, along with energy-positive buildings – also unimagined 10 years ago.[93]

These are hints for healthy disruption, although care has to be taken that concepts such as 'collaborative' and 'sharing' are fully enforced to lower our ecological footprint in an equitable manner, and not abused for the creation of new private monopolies by using digital tech to bypass rules, in particular taxation and labour regulations.

The transition in the philosophy of science from the reductionist methods to more life respecting approaches (Sect. 2.7) can greatly benefit from the availability of ICT methods to simulate complex, evolving, responsive living systems. For the Club of Rome, it has been very encouraging to see the methodological evolution from the simple World3 computer model of the 1972 *Limits to Growth* to Jørgen Randers' *2052*, 40 years later.

Of course, the ICT revolution goes considerably further than information exchange and methods for modelling and understanding complex systems. Our entire industrial sector is currently in an exciting transition to 'Industry 4.0'. In Sect. 1.11, Jeremy Rifkin was mentioned who (with a slightly different way of counting) describes the 'Third Industrial Revolution' naming five 'pillars' characterizing it, mostly relating to renewable energies and their tendency to decentralize power supply and all associated production processes. For developing countries so far lacking complete high voltage mains, this is an exciting chance leap to frog past some damaging development phases.

A different kind of IT-driven progress relates to the availability of information through the Internet and Wikipedia, information that in earlier times was stored in libraries and periodicals, which took far longer to travel to, full days or even weeks to research. Moreover, websites make big and small companies, government offices, foundations and activist groups visible in places that used to be essentially cut off from a wider world.

Then there is the prospect, in part the reality, of IT-based democracy. While already practised for a long time in some places, 'direct democracy' is now becoming more easily extendable, at least technically speaking, with the help of the many social media platforms that are part of the ICT revolution. But as mentioned earlier, some problematic features of social media affect the wisdom of polls, as well as creating our silly opinion silos. Such phenomena, however, should not be used as a principal argument against electronic support for ad hoc democracy.

3.10.3 And Now a Shocking Proposal: The Bit Tax

Adam Smith in *The Wealth of Nations* said that wealth was based on the division of labour and on taxing production factors. This has inspired the Canadian chapter of the Club of Rome in their book *The New Wealth of Nations* to consider taxing the

[93] l.c. pp. 187–198.

new production factor of information. That was 20 years ago; the driving intellect behind the idea was the late T. Ranald Ide, in short Ran Ide. He and his co-authors said, 'The new wealth of nations is found in the trillions of digital bits of information pulsing through global networks. These are the physical/electronic manifestations of the many transactions, conversations, voice and video messages and programs that, taken together record the process of production, distribution and consumption in the new economy'.[94] As a consequence of this observation, the authors proposed levying a tax on digital 'bits'. Such a tax would be extremely small but still large enough to generate fiscal revenue which could be used to combat the negative externalities of ICTs, as well as to fund new designs for sustainable development.

More importantly is the steering effect of such a form of taxation. If you tax energy, the use of energy will generally become more prudent, and technologies that save energy will become more profitable. If you tax human labour, you create an incentive for increasing labour productivity, which has resulted in fewer jobs. If you tax bits, you will frustrate the senders of Spam and of other unwanted information, and please most users. Of course, there are always some downsides, just as there are for the VAT or for labour, energy and property taxes. But the automatic outcry 'you are taxing progress' is complete nonsense. A tiny tax, in the vicinity of perhaps a millionth of a dollar per bit, would not discourage *any* proper use of information, including the advertising that funds the Worldwide Web. In today's context of poorly functioning taxation, Ran Ide's idea should be re-introduced into the political debate. By the way, the idea can also be weighed against Bill Gates' and others' idea of a 'robot tax' to reduce the size of job-killing by robots.

The bit tax cannot overcome the scandalous fact that companies like Airbnb and Uber manage to do billions of dollars of business worldwide while paying almost no tax to anyone. They use tax havens as their company base but their practices reduce tax revenues from companies and individuals who do pay these expenses, often putting them out of business. Similarly and significantly, the IT giant Apple is being required to pay 13 billion Euros in back taxes due in countries where they operate.

Some of the gurus of the information industry, aware of the danger of being perceived as the main cause for new forms of unemployment, have been championing the idea of an unconditional basic income.[95] This concept is definitely an important part of the debate now starting, on how to profit from the impetus of technological disruption in a constructive, purposeful way. It is crucial to address the challenges facing humanity, the old as well as the new ones that are being created by that very disruption. This situation is truly an exciting opportunity society needs to seize with new thinking about both income and taxation!

---

[94] Cordell et al. (1997).

[95] e.g. Jathan Sadowski. 22 June, 2016. Why Silicon Valley is embracing universal basic income.

3.11 Reform of the Financial Sector

Section 1.1.2 dealt with the inherent risks of the monetary system. The system is unstable. It creates asset bubbles. Only a minor part presently supports investments in the real economy and the system drives inequality, increases volatility and tends to work procyclical, that is, amplifying booms and busts. The financial crisis of 2008–2009 demonstrated how damaging these features can be. Lastly but not the least, financial institutions willingly lend out large sums to, or invest in, companies whose risk exposure is significant when energy and climate and other environmental issues are taken into account (see Fig. 3.6) – at great risk for stockholders including pension funds, and worsening pollution of the atmosphere and the destruction of vital ecosystems.

The question is what to do about it? How can the global economy and the monetary system be redesigned so as to respect the principles of sustainability? As authors of 'Come On', we are no experts in the monetary system – far from it. However, extensive reading and discussions with a variety of experts have led us to suggestions outlined in the following. A number of measures have to be considered. A central one would be to address the main driver of 'growth' and financial instability: debt.

Banks, along with Central Banks through their quantitative easing measures, are the main money-makers today. Deregulations during the 1980s and onwards led to a massive increase in money creation. Bank revenues have more than tripled in the OECD countries. The build-up of private and public debt, which is the other side of the coin, is part of a credit-fuelled growth model favoured both by centre-right and centre-left politicians. Such policies have been supported, as well, by most monetary authorities and central banks.

To arrest the expansion of debt, we need increases in compulsory capital reserves and ratios as well as tighter controls on private credit creation.[96] While mainstream economists – and the public – appear to assume that lending is financed primarily by savings, this is manifestly untrue. Banks create money in the act of creating debt – to a large extent out of thin air.[97] There are limits in the form of so-called money multipliers that are linked to the different asset classes that the new money is intended to finance. These money multipliers are under political control and could be used more actively to steer money creation towards the real economy rather than into financial bubbles. But these constraints have become less and less stringent and as a consequence credit has expanded far beyond the needs of the real economy.

The immediate challenge will be to stimulate the banks to create money for real investments rather than for excessive speculation in different types of financial assets and consumer or real estate credits. This being said, it has to be recognized

[96] Turner (2016).

[97] The fraction of a bank's debt that is tied to actual reserves (savings) is extremely low in developed countries (typically only 1 or 2%), having been cut drastically in the past 40 years—from levels closer to 20%.

that it is technically difficult to stifle the creation of money for purely speculative purposes. The incentives to find ways around any barrier imposed by policymakers are significant.

No doubt, we are in a race against time. Excessive money creation and the debt addiction must be curbed to stabilize the system. On the other hand, if lending were to be cut off too abruptly, a 'liquidity drought' would immediately develop and asset bubbles might burst and many banks fail.

Ulf Dahlsten, former state secretary in the Swedish Government, describes the challenges the following way in a forthcoming book[98]:

'The major problem is that the financial markets are increasingly global, while the supply of institutions being supplied are mainly national. There is a lack of institutions on the global level to make it possible to decide upon and implement laws and regulations in the common interest. There is no global lender of last resort, no central bank to manage global imbalances, global liquidity, reserve currency issues, international regulations, resolutions and the like on a regular basis. There is the IMF, which could take on these tasks, and there is extensive global networking, but no international decision making power. The financial markets are probably the area in which the need for an International Market Law is the most obvious. International institutions, central bankers and regulators all need new authorities and new and reinvented tools'.

A number of measures surely deserve serious consideration for reforming the financial sector.

3.11.1 Separate Commercial and Investment Banking

The separation of commercial banks from investment banks provided financial stability for more than 40 years after 1933. If commercial and retail banking were to be separated again from speculative investment banking, taxpayers would no longer be called upon to bail out banks destroyed by speculative failures. Such banks would cease to be tied to citizen deposits that warrant government protection. Proposals along these lines have been made, both in the United States and the EU. But so far limited action has been taken. The US Congress did decide on some measures during the Obama administration, but after the election of Donald Trump and with the Republicans in control of both houses efforts are under way to lessen regulation of financial markets rather than tightening them.

A further step could be to turn back the trend towards ever bigger commercial banks turning their backs on client groups like start-ups and family businesses which are less profitable but nevertheless very important for our society. This could happen by granting limited banking licences for specialist banks focusing on certain client types or regional areas. Lawmakers and regulators should be more careful and selective in granting the privilege of a banking licence, in relation to both business

[98] From manuscript for a forthcoming book; title and publisher not yet available.

purpose and geographical area. The trend over the last 50 years has been in the opposite direction. Fewer and bigger commercial banks are more profitable for the shareholders of those banks, but they are often less efficient in serving the needs of our societies.

3.11.2 Dealing with Debt

Increasing the amount of debt in society is not necessarily a problem in itself. It could be viewed as a healthy sign of an increase of the trust level in society. More money in the system makes it possible for more things to happen: new businesses started, new technologies developed, infrastructure created and more people potentially lifted out of poverty and need. The problem is *how* this newly created money is used.

However created, debt is a claim on the actual wealth of the future. As Australian economist Richard Sanders explains, 'In the simplest analysis, the root of the sustainability problem is an exponentially growing set of claims (money) on a finite (and indeed diminishing) pool of natural capital'.[99] As well as this mismatch with ecological realities, excessive credit leads to financial crisis and leaves behind a crippling debt overhang, such as exists today.

What has been very much lacking is sufficient loss absorption capacity in banks on account of their excessive debts, with a leverage ratio between 3% and 5% being the norm for the largest of the banks. What that means is that most banks can still absorb only about 5% losses to their balance sheet before going bankrupt. Efforts are being made to tighten the leverage ratio. The new capital adequacy rules anticipated by Basel IV would no doubt strengthen the solvency of the banking system. According to McKinsey, the implementation of the rules proposed would increase the average capital ratio for European banks to 13.4%.[100] This would, no doubt, be a step in the right direction. However, we would rather see a quadrupling of the leverage ratio. This would go a long way towards stabilizing the system and protecting taxpayers. Whether the changes will go far enough is difficult to say. However, we would like to see substantial increases in the leverage ratio. We refer again to Anat Admati and Martin Hellwig: 'Banks usually ask us to hold 20 percent equity when we borrow. We should ask the same of them'.[101]

Savings banks and local banks tend to argue that they know their customers well enough to assess the credit risks. They also want to avoid bureaucratic and leverage cost resulting from global regulation that is necessary for global players but could kill locals. It is not up to the Club of Rome to judge but we suggest that reforms allow for some differentiation between local and global institutions.

[99] Sanders (2006).

[100] McKinsey (2017).

[101] Admati and Hellwig (2013).

Dealing with the vast pool of existing debt is a difficult problem. Whether the debtors are countries from the global South, or are countries such as Greece, Spain and Italy, the owners of much of the debt are the large international banks and other players in the capital market who resist any substantial forgiveness of debt, even though some of it was quite predatory at inception and in fact not repayable. Some of the debt and the non-performing loans should be written off, at least to repayable levels. And the rest should be moved progressively out of bank balance sheets so as to break the vicious sovereign bank nexus. In some instances, bank balance sheets remain so full of sovereign debt that debt restructuring of sovereigns may lead to a potential banking crisis.

3.11.3 Control Money Creation: The Chicago Plan

The money in circulation today is, according to most experts, several times larger than what is required to support the real economy. This is largely the result of unchecked money creation by the banks. The proposals discussed above, that is, to tighten the leverage ratio, would no doubt help remedy the situation. If banks had to hold total capital funds in the range of 20% of their liabilities, the nature of their business would have to change. No longer would they be ready to take on risks such as they have in the recent past.

But many people think that the tightening up of leverage ratios and the mobilization of more equity is not sufficient to reform the system. They compare the situation today in terms of indebtedness and money creation with the situation at the time of the Great Depression in the 1920s and 1930s. A radical approach to address the problem of too much debt at the time, which later became known as the Chicago Plan, was proposed by a Nobel Prize Winner in Chemistry, Frederick Soddy in 1926. *He argued that the creation of money should return exclusively to the state.* The Chicago Plan was developed in the 1930s, championed primarily by Professor Irving Fisher of Yale University. However, it failed to gather enough support within the Roosevelt Administration, which chose to back stronger regulations of the banks instead.

In the aftermath of the 2008 financial crisis, the Chicago Plan has received renewed attention. Several NGOs have looked into the plan and endorsed it. The think-tank Positive Money has developed a detailed proposal for reform of the British banking system based on the Chicago Plan.[102]

Of particular interest is a recent scrutiny of the Chicago Plan performed by the IMF economists Benes and Kumhof.[103] They have used an up-to-date model of the US economy to verify Fisher's findings and have found support for all his claims. By taking over the role of creating money, the state would cover all deposits in the banks and cancel any risk of bank runs. State debt would be reduced by 40% and the average household become debt-free. According to Benes and Kumhof, there would be 'no losers'.

[102] Tekelova (2012).

[103] Benes and Kumhof (2012)

3.11.4 Tax Financial Trading

A small 'Tobin tax' on financial transactions (perhaps from US$ 1 m upwards) should preferably be implemented worldwide. But past negotiations show that this will not happen in the near future. It is more realistic to let some strong countries start with the tax, discouraging speculation, and let the pioneer countries keep the revenues for themselves.

3.11.5 Enhance Transparency

The entire derivatives market should be scrutinized to see if any of it serves any purpose other than speculation. Purely speculative instruments could be phased out or taxed, and any derivatives that are regarded as useful should be moved into the plain sight of properly regulated and globally supervised centralized counterparties. The 'shadow banking system' (approximately 70% of all banking at the time of the 2008 crash) should be curtailed by regulation as tight as that applied to banks themselves.

3.11.6 Independent Regulators

Regulators are usually drawn from the class of banking executives who run the transnational banks. Regulators need to be genuinely independent; extended 'waiting periods' could be imposed to assist this objective. Bankers don't mind regulation per se but tend to oppose overregulation. Regulators should respect the 'principle of proportionality'. That means the regulation that is essential for large banks could be somewhat softer for small- and medium-size banks.

3.11.7 Taxing the Rich and Collecting the Tax

The combination of tax avoidance, tax evasion and secrecy jurisdictions where assets can be hidden (tax havens) is a system that, as well as facilitating money-laundering for criminals and dictators, can also insulate more legitimate wealth from its social and financial obligations.[104] It is estimated that, as of 2012, between US$ 21 trillion and US$ 32 trillion was hidden in these secrecy jurisdictions.[105]

[104] Shaxson (2012).

[105] Henry (2012).

Their essential features are low or zero tax and secrecy provided through a maze of shell companies where beneficial ownership is impossible to identify.

Efforts to extract taxes from TNCs and high net worth individuals (HNWIs) will need international cooperation and serious application. Representatives of tax-avoiding corporations always state that they are complying with all laws. In many cases, this is true, so laws must be changed. As a rule, companies should pay taxes in the country where their profits are made.

Bringing about full transparency is the first requirement. Oxfam calls for a public registry of ultimate beneficial owners of companies, foundations, trusts and accounts, including those held in secrecy jurisdictions. The OECD has been edging towards reform for 20 years and is beginning to implement its automatic exchange of information (between governments, including all banking data of their residents) and its country-by-country reporting standards.[106] When these are in place, expected during 2017 and 2018, it is thought that some of the tax avoidance practices that are being used today will no longer be possible.

Economist Gabriel Zucman, however, warns that progress has been glacial and of little effect and that tax accountants and lawyers, being far better resourced than tax departments, may well circumvent these rules – wealth hidden in tax havens increased by 25% in the past 5 years, despite OECD action. Zucman proposes devising a formula where a TNC's total global profit is apportioned to the countries where it is earned, nullifying avoidance schemes.[107]

This entire system contributes significantly to rising inequality. At present, developed countries lose revenues that should have been applied to health, education, environmental protection or other national priorities. Their prevailing budget deficits would all be less serious if these taxes had been collected. For the global South, the loss is even more drastic, since there is little pre-existing infrastructure for the welfare of citizens and inadequate capital to finance sustainable development.

Oxfam has called for the creation of a global tax body, tasked with assessing the risks posed by secret jurisdictions. Oxfam recommends that the outcomes be made public as a disincentive for those who use and facilitate the use of tax havens. Oxfam favours cooperation with the IMF and OECD to develop a list of tax havens, so that governments can legislate disincentives.[108]

3.11.8 Curbing the 'Big Four' Accounting Firms

The historical role of accounting firms was to audit and verify accounts of corporations. After the financialization era (Sect. 1.1.2), only five big firms were left: PwC, Deloitte, KPMG, Ernst & Young (EY) and Arthur Andersen which collapsed in

[106] OECD (2016).

[107] Zucman (2015).

[108] Jamaldeen (2016).

2002.[109] The remaining four giants, sheltered by opaque partnership structures, are auditing 98% of corporations with a turnover of US$ 1 billion or more. They have also been assisting numerous large corporations in developing tax avoidance schemes, such as the Luxembourg schemes leaked in 2014.[110]

Such schemes cost governments and their taxpayers more than US$ 1 trillion a year, according to tax lawyer George Rozvany,[111] an insider who has worked for EY, PwC and Andersen. He recommends that the accounting/audit businesses be separated from the consultant/tax advisor businesses in all four firms, a similar reform to the separation of commercial and investment banking.[112]

Such recommendations may hinge on governments regaining power over the fate of the people they represent and the ability to make the key choices about direction and objectives. They also depend on the OECD and the IMF to implement and *enforce* proposed new rules. As was said in Sects. 2.6.1 and 2.10.3, we need a healthy balance of powers between private and public interests, that is, between private business and the state or the international bodies representing the states of the world.

Significant political will is needed, countering the capture of many governments by vested interests that will resist change. It is unfortunate that financiers were able to navigate through the actual crisis of 2007–2008 with little intrusion on their industry, despite the fact that taxpayers propped them up with vast sums. An opportunity to insist on reversion of control to sovereign governments was squandered by representatives under the influence of the finance industry (especially in the United States). It is crucial that governments avoid such an outcome if and when the next crisis arises.

3.12 Reform of the Economic Set-Up

The reforms of the financial sector have to be discussed rather urgently, and they are, in fact, on the agenda inside states and at the international level. But in a sense, they are only a first step. For a true and lasting change for the better, it will be necessary to go deeper and look at the economic and political systems that have formed the current state of the financial sector.

One striking example, currently under stress, is the EU. It emerged as a common market, comprising six countries that had experienced almost continuous wars against each other for more than a thousand years. Economic integration brought the six together in a peaceful way, making it totally absurd to even think of fighting wars. Beginning with coal and steel, soon adding agriculture and eventually covering all relevant economic sectors, attracting more and more countries and creating

[109] Brown and Dugan (2002).

[110] Bowers (2014).

[111] West (2016a).

[112] See also West (2016b).

common infrastructures, the EU became one of the greatest success stories of the post-World War II history. The EU also represents a sense of balance: the balance between local, provincial and national functions on one side, and European functions that are better served if done at Community level. Schools are local and national, but recognition of graduation certificates is European, obliging states, however, to agree on quality standards.

The ecological crisis discussed chiefly in Chap. 1 will force us to go much deeper in reconsidering economic structures. New thoughts are needed to deal with the twenty-first-century challenges. Also other challenges require new thinking such as the disappearance of many job functions, the demographic transition, the scandals of injustice and the scourge of terrorism. In this chapter, only four examples taken somewhat at random that respond to these challenges are presented in some detail.

3.12.1 'Doughnut Economics'

Oxford economist Kate Raworth, member of the Club of Rome, has written a book called *Doughnut Economics*[113] in which she argues that the prevailing ideas about economics are centuries out of date. Today's students – the policymakers of 2050 – are being taught ideas from the textbooks of 1950 which, in turn, are grounded in the theories of 1850 or even earlier. Raworth says that given the challenges of the twenty-first century – from climate change to extreme inequality to recurring financial crises – this dinosaur approach is shaping up to be a disaster.[114]

Raworth frames the challenges we face in a new way. Humanity's goal this century is to *meet the needs of all systems and living beings, within the means of the planet;* it can be pictured as being shaped like a doughnut (the kind with a hole in the middle), with outer and inner boundaries. The planetary boundaries, according to Rockström (Sect. 1.3), are the outer, limiting ring; and a set of social challenges, much along the lines of the SDG agenda (Sect. 1.10), are the inner ring.

We need to contextualize our economic thinking, and she points out that context brings meaning. The economy is not like a machine that operates *on* the world, but is something embedded *within* it, more like the heart and its circulatory system. In this respect, Raworth makes use of insights from *systems* science.

Her book outlines seven principles to guide economists in developing the institutions and policies needed to bring humanity into the Doughnut's safe and just space between social and planetary boundaries. Seen this way, the economy becomes the toolbox that helps make both equality and sustainability happen. This is more or less the opposite of the rules guiding our economies today, where maximum growth of the economy is the overriding and constant goal, and social and ecological objectives are addressed as afterthoughts.

---

[113] Raworth (2017)

[114] Kate Raworth in a blog post on April 7, 2017, the day of the launch of her book.

Most of Raworth's seven principles are already part of the sustainability debate. They may seem altruistic, and one can maintain that individuals or groups will first and foremost think of their own advantage. However, she argues that under proper social guidance, the (higher) advantage of the community can, and should, become more attractive:

1. *From GDP growth to the Doughnut*: The purpose of the economy is far more than pursuing GDP growth; it is to *meet the needs of all within the means of the planet*. This single switch of purpose transforms the meaning and shape of economic progress – from endless growth to thriving in balance.
2. *From self-contained market to embedded economy*: This point aims to state the obvious that the economy is embedded within society and within the living world, and is dependent upon and in service to them both.
3. *From rational economic man to social adaptable humans*: Human nature is far richer than the individualistic selfishness of current economic theory suggests. We are reciprocating, interdependent, approximating, adaptable, social beings embedded within the web of life.
4. *From mechanical equilibrium to dynamic complexity*: Newtonian physics is the wrong kind of science for economic analysis; it is far smarter to embrace the complexity and evolutionary thinking of systems science.
5. *From 'growth will even things up again' to distributive by design*: Trickle-down economics doesn't work. It is time to create economies in which value generated is shared far more equitably with all those who helped to create it.
6. *From 'growth will clean it up again' to regenerative by design*: It is a myth that economic growth will – like a well-trained child – clean up after itself. Regenerative, circular design is far more elegant than degrading then attempting to restore the living world.
7. *From growth-addicted to growth-agnostic*: Today's economies need to grow whether or not they make us thrive. Tomorrow's economies must be able to make us thrive whether or not they grow.

3.12.2 Reforms that May Find Majority Support

In their book *Reinventing Prosperity*, Maxton and Randers[115] make the point that all the major challenges facing human well-being today, like widening income inequality, continuing global poverty, and environmental degradation, are relatively easy to solve – in theory. In practice, however, they are much more complex, because most of the commonly proposed 'solutions' are simply not acceptable to people and governments focused on the short term. In interviews, Randers is quoted as saying that to solve these kinds of problems in a democracy strikes him as being very difficult, if even possible. In contrast, he depicts China as an example of a nation that is able

[115] Maxton and Randers (2016).

to deal with long-term issues and which implements policies that deals effectively both with poverty reduction and climate mitigation.

The authors go on to make a serious attempt to present policy measures and solutions that would address the main challenges humanity faces, solutions that would be possible to implement around the world. Their book addresses the 40-year-old growth/no-growth debate by explaining how it is possible to reduce unemployment, poverty and inequality and effectively address climate change and ecosystem decline and *still have economic growth* – if so desired. Maxton and Randers identify which 13 proposals relate to these systemic problems of unemployment, inequality and climate change.

Reducing the scourge of unemployment can, at least mathematically, be achieved by shortening the average work year. It means redistributing work to more people who are willing to work. Employed people would have more time for their families, recreation, acquiring new knowledge and skills and a general widening of the horizon. The average productivity per workhour need not suffer. Incumbent workers and their employers may not like the idea, but they can live with it, notably in countries visibly suffering from high unemployment. Programmes could be introduced in small increments, allowing for learning and adaptation.

On the other hand, it may be reasonable to allow people to work as long as employers or clients are satisfied with the work they do, that is, by raising the retirement age. If longer work life becomes customary, it can be expected that functions, or jobs, will increasingly develop that are particularly suitable for persons with long experience in moderation, coordination, and conflict resolution but who may suffer from the typical physical weaknesses of the elderly. Also functions requiring acquaintance with old-fashioned tools or products – for repair and maintenance – can be suitable for older employees or freelance workers. Inasmuch as working seniors reduce the cost of the welfare state, the state can spend more money on new job creation activities.

A third idea in this context is remuneration for caring for others at home. Riane Eisler calls for a 'Caring Revolution' consisting of a major upgrade, financially and in terms of reputation, of caring work.[116] To make such work more visible, a formal organization may be required looking after acceptable payment and working conditions. Remuneration should primarily come from public budgets so that the total cost is shared by all taxpayers. Public acceptance would grow as people see it as an adequate response to the challenge of an ageing population.

One step further is, as mentioned, the idea of an unconditional income for all, that is, not conditional to specific work. It is currently discussed in many quarters, not least in Silicon Valley in companies realizing that their huge profits are a result of destroying millions of jobs.[117] It is worth initiating a broad discussion on this so far controversial proposal, both on the national and international levels.

[116] Eisler (2007).

[117] We quoted already Jathan Sadowski. 22 June, 2016. Why Silicon Valley is embracing universal basic income.

A related idea is to increase and guarantee higher unemployment benefits. Other measures of reducing unemployment and thereby inequality can fall in the category of public stimulus packages. These may include physical infrastructures, lifelong education and environmental restoration.

Initiated by World Bank President Jim Yong Kim, a Carbon Pricing Leadership Coalition (CPLC) was formed during the COP 21 climate meeting in Paris (see Sects. 1.5 and 3.7). The idea – not new, of course – is supported by Maxton and Randers, and aims at taxing all fossil fuels. The novelty is that the proceeds would be distributed equally to all citizens. This benefits the majority, the poor and those with lower energy consumption. At the same time, it encourages the imperative shift to clean energy.

The same idea can be generalized into a shift of taxation from human labour to the use of the physical resources of the planet. This would also shift commercial interest from labour productivity to resource productivity (see also Sect. 3.9).

It can be very popular among democratic majorities to increase inheritance taxes. Again this will require some international harmonization. The state can devote the proceeds to agreed social priorities but can also leave many of the choices to ageing individuals.

Another international task is reforming the WTO rules allowing countries to impose tariffs on products and services that clearly damage the environment. As said in Sect. 1.9, WTO rules so far have a strong anti-environment bias. These last two points relate to the broader problem of global governance that we address in Sect. 3.16.

3.12.3 Making the Green Transition Ever More Profitable

Some of the exciting initiatives sketched out in Chap. 3 show that much can be done even under today's conditions. But the success stories are the exceptions. The main stream in all countries is still conventional, often destructive to the environment, often favouring the rich and disfavouring the poor. If the transition to a sustainable world society is ever to become mainstream, policy measures are required to make sustainable businesses more profitable than non-sustainable businesses.

Conventional environmental and social policy measures have been command and control instruments for the environment and taxation for social equity. By and large, both have improved the state of affairs in terms of polluted air or water and social justice. But they failed to correct the unsustainable trends described in Chap. 1 of this report.

The most effective transition method for a sustainable society is *changing the financial frame conditions for business*. Of course, mandatory standards and bans must continue, like in cases of pollutants and risky chemicals, most urgently in agriculture. However, standards will only marginally curb energy consumption and greenhouse gas emissions. For massive progress in this regard, economic instruments are unavoidable, putting price tags on GHG emissions, on energy, water,

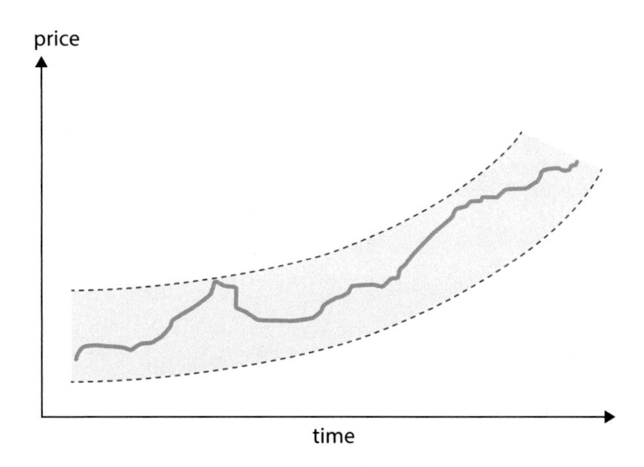

Fig. 3.12 To minimize state interventions, a corridor can be politically agreed (*dotted lines*). If market prices (*blue line*) hit the upper or lower corridor borders, the state would intervene to bring it back into the corridor

primary minerals and soil degradation. The easiest way of doing this, is taxes. They would go up if world market prices fall, and could be reduced if market prices rise. The aim is high predictability.

Taxation of energy, water, etc. is easier said than done, because it will be extremely unpopular. So the main political challenge is finding pricing policies that find political majorities.

One method has been outlined in UNEP's *Decoupling 2.*[118] It suggests increasing prices for energy and other resources *in parallel with empirical advances in resource productivity*. If the average energy productivity, for example, in private households, increases by 1% in 1 year, the prices for energy in households would rise in the following year by 1% (plus inflation). Similar for transport, industry and services. If everyone knows that prices will go up in this way, we can expect a self-accelerating dynamic, because it would become ever more profitable, year by year, to invest in added energy efficiency.

In order to avoid too many interventions, a *corridor* could be politically agreed (dotted lines in Fig. 3.12). Market prices may fluctuate (blue line). But if market prices hit a corridor line, a corrective intervention brings it back into the corridor, thus discouraging speculation.

Consumers, producers, traders, engineers and investors will pay ever more attention to resource efficiency. If a fivefold, in some cases 20-fold, increase in energy productivity is technically achievable (Sect. 3.9), one could expect improvements to become ever more impressive.

There are a number of problems to be addressed:

- Existing industry processes such as melting aluminium from bauxite, or electrolysis anyway, have almost exhausted their potential for improvement of energy efficiency. But in many cases, substitutions of methods or of materials can still lead to jumps in efficiency.

[118] International Resource Panel. 2014. Decoupling 2. Technologies, Opportunities and Policy Options. Nairobi: UNEP, see also Factor Five, l.c. Chapter 9: A long-term ecological tax reform.

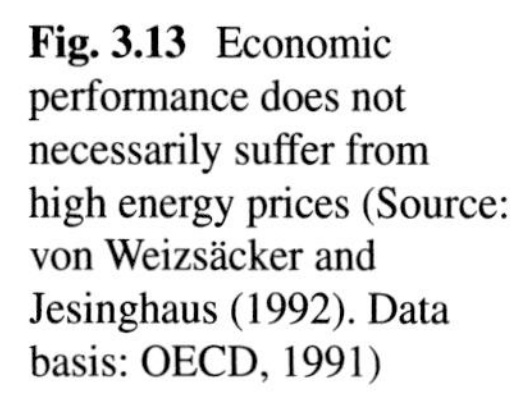

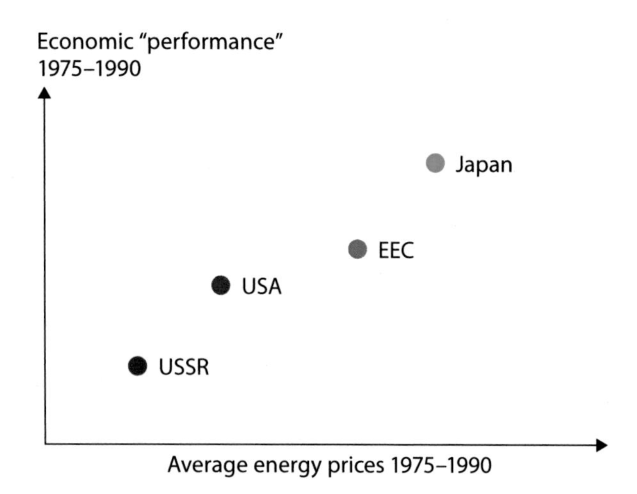

Fig. 3.13 Economic performance does not necessarily suffer from high energy prices (Source: von Weizsäcker and Jesinghaus (1992). Data basis: OECD, 1991)

- Technological progress reaches poor families a lot later than the well-to-do. It would therefore punish poor people if rich people become more efficient and thereby make prices rise, according to the suggested policy. The answer can be low 'life line tariffs' that are not affected by the said policy.

Certain industries simply cannot compete internationally if their domestic energy prices grow higher than those abroad. The answer would be twofold: try and harmonize the policy internationally; and as long as that isn't properly working, the energy tax revenues collected from such industries can be returned back to the industry, based on added value or number of jobs. The respective industrial sectors would not lose money, but inside the sectors, there would be a strong incentive for enhanced energy efficiency. Sweden in 1992 had such a 'revenue neutral' scheme when introducing a hefty NOx charge applicable to stationary combustion plants, chemical industry, waste incineration, metal manufacturing, pulp and paper, food and wood industry. Industry was happy with the challenge because the respective industry sectors lost no money. In the end, they even became more competitive.[119]

The fear of losing competitiveness should not be exaggerated. During the 'energy crisis' of the 1970s and 1980s, four different economic regions maintained different strategies on energy prices. In the Soviet Union, prices were and remained low despite high OPEC oil prices. In the United States, governments feared that the 'American Way of Life' would suffer too much if gasoline prices were high, so essentially no fuel taxes existed, and the United States still had a lot of domestic oil. In Western Europe, petrol taxes remained high, and electricity was expensive. Japan, finally, having no domestic fuels to speak of, and being plagued by air pollution, had very high petrol taxes, and electricity was very expensive, in part due to pollution control measures, and also to finance the build-up of nuclear power. What happened to the economic performance of these four regions? Figure 3.13 shows the stunning answer: The higher the energy prices, the more successful was the region at the time!

[119] Höglund-Isaksson and Sterner (2009).

OK, the causality may be the other way round. But in Japan, historically, the big success happened *after* energy prices exploded. Clearly, there are hundreds of other measures that can help making the transition to sustainable development more profitable, but in the absence of a pricing regime, they would involve mega bureaucracy and are unlikely to change the big picture.

3.12.4 Economy for the Common Good[120]

A finer future characterized by the concept of a regenerative economy is a wonderful vision and should be part of any progressive political programme. That could be supported by practical innovations occurring already. One concept by a young Austrian holistic thinker and author, Christian Felber, is termed the 'Economy for the Common Good' (ECG). He thought there should be ways and means of avoiding the 'Tragedy of the Commons',[121] by redefining the commercial objectives of private companies. Conserving the common good would become a new ethical imperative for businesses subscribing to his ECG.

Reflecting on the causes of the countless collateral damages of the current economic system, Felber and his friends, chiefly entrepreneurs and business consultants associated with the 'Attac' movement,[122] formulated the ECG over a 2-year process. It included a 'common good balance sheet' which was presented at a public meeting in Vienna in October 2010. In the course of the meeting, 25 companies volunteered to implement this innovative balance sheet in their next fiscal year, and the conference participants decided to initiate a movement for the implementation of ECG.

ECG is meant to encourage and enable businesses to transform the conventional profit maximization principle into an orientation focused on the common good, while transforming the principle of competition into a principle of cooperation. This programme incorporated the insights of Alfie Kohn, Joachim Bauer, Gerald Hüther and Martin Nowak, to name a few.[123]

One philosophical branch of the ECG goes back to Aristotle's Nicomachean Ethics, which says that when we aim at happiness, we do so for its own sake, not because happiness helps us realize some other (including financial) end. Starting with an objective of changing the overall incentive structure that underlies neo-liberal economic thinking, the founders of the ECG focused on identifying those values that should in the future become prominent, according to which this

---

[120] We wish to thank Volker Jäger for a draft in German of this sub-chapter and Bodo E. Steiner for translating it into English. Editing responsibility rests with the book's lead-authors and the editor.

[121] Garrett Hardin (1968).

[122] Attac is a social movement, active chiefly in Europe, and critical of neo-liberal globalization. The name came from the French Association pour la taxation des transactions financières pour l'aide aux citoyens.

[123] E.g. Kohn (1990).

orientation towards the common good is to be achieved. These values were chosen from a set that underlies almost all democratic constitutions across the globe, namely:

- Human dignity
- Solidarity
- Ecological sustainability
- Justice
- Democracy (transparency and participation)

Felber argues that if we all let interpersonal and ecological relationships succeed and develop, society would discover a new meaning for performance and economic success.[124] A new economy of connectedness could emerge that could provide an alternative to the current economy of rivalry and separation.[125]

A practical manual for the Economy of the Common Good[126] allows companies and external evaluators to arrive at more or less consensual results in their operations and evaluations. Currently, there are about 400 companies in Germany, Spain and Austria that have been assessed and audited voluntarily, achieving between 200 and 800 'Common Good' points.

A next step could be for governments to honour high scores by granting tax privileges, favourable loan conditions, or purchasing preferences for public procurements. This is the political aim of the international ECG movement, now comprising over 3000 volunteers in 150 local chapters.[127] The first municipalities and regional parliaments have already announced they will be giving companies with good balance sheet results priority in terms of public procurement. Even as these legal incentives start to enter into force and provide economic rewards for ethical business strategies, the pioneer companies are already reporting additional advantages, which include improved customer ties, staff loyalty and reputational gains.

Out of the more than a hundred companies adopting ECG, only one will be presented here, and it is from perhaps the most unlikely sector of our economy, that is, banking. This case study is the savings bank of Dornbirn, Dornbirner Sparkasse, Vorarlberg, Austria. It belongs to a system of Austrian savings banks founded in 1819 by Father Johann Baptist Weber.[128] Its initial objective was to make banking accessible to a broader population base. Weber, not unlike Bangladesh's Muhammad Yunus of the present day, believed that if there could be savings banks, seeking ordinary and rather poor people as its clients, the whole population would benefit.

---

[124] Felber, Gemeinwohl-Ökonomie 2012 (German edition): p. 107.

[125] Eisenstein, Charles: Ökonomie der Verbundenheit, Scorpio, Berlin, Munich, 2013. In Englisch: http://sacred-economics.com/about-the-book/

[126] Publicly available Document under Creative Commons for companies and auditors for self-assessment and the assessment of other companies, according to the criteria of the CGE Matrix. English version is here: https://old.ecogood.org/en/download/file/fid/556

[127] Felber (2012, p. 47).

[128] https://de.wikipedia.org/wiki/Erste_Bank

The Dornbirner Sparkasse AG was established in 1867 as a community bank by the town of Dornbirn (ca 50,000 inhabitants) and was converted into a public limited company in 2002. The shares are now owned by a bank entity in charge of administering the shares and by the town of Dornbirn. The bank has 14 branches in the small province of Vorarlberg. It has approximately 350 employees and has a balance sheet of 2.3 billion Euros (2015). At 18.3% (2015), the equity ratio was well above that of many of the larger business banks.

The Dornbirner Sparkasse adapted its projected strategy for 2020 with intensive employee involvement. It takes values-based thinking and customer-centred orientation as a means to become close to its clients. These values include using banking as a means of practicing and expanding appreciation, openness, trust, determination, courage and sustainability. During the process of strategy planning and in the context of 'common good accounting', these values have been deepened and the bank has developed even more binding commitments to them. They also decided to have them audited and certified externally. During the 2016 audit, the core team confirmed that the internal discussions during the auditing process had triggered a great deal of action towards the common good orientation. The bank is proud that in the process of establishing the common good balance sheet it experienced a renaissance of the original values of the Austrian savings bank.

Another activity relevant for receiving high ECG scores that interested the Dornbirner Sparkasse Bank and its employees is ethical financial management. Assets with poor ethical ratings were phased out from the bank's portfolios and replaced by 'common good'-oriented ones. This process sharpened the participants' views on ethical investment and its link to the bank's daily operations. What is perhaps most important is their bottom line has not suffered in the least from the new orientation. On the contrary, they have been able to attract a number of business clients for this very reason, including a cooperative with almost 5000 members, which opened its business account with the Dornbirner Sparkasse and also deposited part of its equity there.

3.13 Benign Investment

Traditional investing is being practised and taught in business schools in terms of how to use money to purchase a financial product, shares in a company or other valuables such as real estate, commodities or art. The intention of traditional investing is to make a financial profit.[129] This purely monetary profit has today become the sole measurement criterion of investment success.

Somehow, our society has come to value cash higher than equity, social impacts or the environment. Moreover, through the recent quantitative easing, central banks have been empowered to print enormous amounts of money to fuel the economy.

[129] Mankiw (1998).

However, much 'cash is rotting away in safes', as one corporate officer put it, while the cash value of some tech companies is going through the roof. For example, some tech start-ups have achieved 'unicorn' status with valuations higher than $1 billion in private markets.[130] As of the end of 2015, 146 private tech companies were valued at more than twice the valuation number of 2014; and 14 private companies enjoyed valuations exceeding $10 billion and are called 'decacorns'. Some got huge public visibility, notably *WhatsApp*, a messaging service firm with annual revenues around $20 million, acquired by Facebook in 2014 for $19 billion, which exceeds Iceland's Gross Domestic Product (GDP) for the same year.[131]

3.13.1 From Wall Street to Philanthropy

On the opposite side of for-profit-only investment spectrum is philanthropy, which has traditionally aimed at balancing out social injustice and environmental degradation with altruistic concern the endowment of institutions and the generosity of people. According to Giving USA, Americans, who have created an extraordinary volunteer-driven culture, donated an estimated $358.38 billion to charity in 2014.[132] The largest portion of this amount comes from individual donations ($258.51 billion); followed by foundation giving ($53.97 billion); bequest giving ($28.13 billion); and corporate giving ($17.77 billion). Approximately a third of that amount went to religious causes, followed by education, human services, health, arts, the environment and social causes. However, these kinds of donations are mostly focused on addressing *national* needs. Few contributions consider global issues and challenges.

The Giving Pledge, launched in 2010, is another example. It's the response of forty of the wealthiest families and individuals in the United States to humanity's quest for solutions to today's challenges.[133] This effort is honourable and greatly needed, and yet philanthropy struggles with its own issues. Today's philanthropic landscape is plagued by outdated legal and management structures.[134] The main reason is that yearly only approximately 5% of the endowment of most philanthropic organizations is programme related, that is, dedicated to the philanthropic mission, and managed by programme managers. Ninety-five per cent of the endow-

---

[130] Erdogan et al. (2016).

[131] The Gross Domestic Product (GDP) in Iceland was worth 17.04 billion US dollars in 2014 viewed on May 11th, 2016 at http://www.tradingeconomics.com/iceland/gdp

[132] Giving USA: Americans Donated an Estimated $358.38 Billion to Charity in 2014; (June 29, 2015). Website viewed on May 13, 2016 at http://givingusa.org/giving-usa-2015-press-release-giving-usa-americans-donated-an-estimated-358-38-billion-to-charity-in-2014-highest-total-in-reports-60-year-history/

[133] Pledge (2010).

[134] Fulton et al. (2010).

ment assets are usually managed by an independent legal entity (often a trust), mostly obliged to preserve and increase it over time unless otherwise stipulated by the founders. Asset managers typically get measured by their financial success, not by the success of the philanthropic mission. Therefore, the largest part of the philanthropic capital turns into regular capital, often invested in companies that make products and services that work against the philanthropic mission. Without major legal changes and structural modifications, the current paradoxes between making charitable contributions to make the world a better place and making investments that actually hurt it, will continue.

Such is the case with the Bill and Melinda Gates Foundation, the second most generous philanthropists in America, after Warren Buffett. While its founders authentically care about their impact in the world and have pledged to spend all of their resources within 50 years after their deaths, the Gates Foundation has been widely criticized for the mismatch between that philanthropic mission and the investments made by their Foundation trust, which clearly serves only to maximize their return on investment. According to Piller et al.,[135] in 2007 the Gates Foundation Trust invested in pharmaceutical companies that price drugs beyond the reach of poor patients and held major assets in terrible polluters. They also purchased 500.000 shares (or approximately $23.1 million) of Monsanto,[136] known for its contempt for the interests and well-being of small farmers and an bad environmental track record. There are, of course, many more such examples of less visible philanthropic foundations.

Very few of currently available financial products contain long-term sustainability characteristics – because those are currently tagged as 'externalities', that is, not worthy of any financial consideration. In Sect. 1.1.2, Lietaer et al. were quoted, showing that some 98% of international financial transactions are essentially speculative, as they are not used for paying for goods and services. And speculation typically has the very short time horizons that we have shown can be so destructive to our collective future.

3.13.2 Ongoing Structural Changes

In an attempt to address the barriers to a sustainable financial future, the association governing the United Nations Principles of Responsible Investing (UN PRI) summarized the key areas in which transformation must be encouraged and future misalignments prevented[137]: short-termism, negligence of environmental and social criteria, lack of transparency and inattention to related externalities.

---

[135] Piller et al. (2007).

[136] GuruFocus (2010).

[137] UN PRI (2013).

The global implementation of such initiatives remains a Gargantuan task, which naturally leaves many questions unanswered. Fortunately, however, transformative developments in the investing industry are mushrooming and there are hosts of investor initiatives geared at integral sustainability, some of which are highlighted below. Part of the motive, of course, is regaining the public trust after the financial industry's failures in recent years.

During the Rio + 20 Earth Summit in Rio de Janeiro in 2012, for example, 745 voluntary commitments were made, 200 of which came from the business and financial community. One is the Natural Capital Declaration (NCD),[138] a finance sector initiative that was co-convened in 2012 during that Summit by the UNEP Finance Initiative and the Global Canopy Programme. It was endorsed by the CEOs of 42 banks, investment funds, and insurance companies. It is aimed at reaffirming the importance of *natural capital* such as soil, air, water, flora and fauna. Its goal is to integrate natural capital in all investment considerations and decisions, as well as in products and services (loans, equity, fixed income and insurance products), along with the supply chains of its signatories.

Moreover, because natural capital is mostly considered as a free good available to all humanity, the signatories requested governments to act quickly to 'develop clear, credible, and long-term policy frameworks that support and incentivize organizations – including financial institutions – to value and report on their use of natural capital and thereby work towards internalizing environmental costs'. According to Thomas Piketty, the real public debt we have is the debt to our natural capital,[139] and the usual assessments like the Gross Domestic Product (GDP) are blind to this danger, as is the rigged profitability of most economic sectors, as Fig. 1.11 illustrates. The fact that the GDP does not integrate natural capital in any way is dangerously wrong and is the main reason for the *Beyond GDP* initiative of the European Commission and other parallel attempts to rectify the measurement of progress (see Sect. 3.13).

The private sector and the investment community have begun to acknowledge that their long-term gains depend on rectifying the meaning of the word 'progress'. The Global Alliance for Banking on Values (GABV), an independent network of around 30 of the world's leading value-based and sustainable financial institutions, has adopted *Principles of Sustainable Banking*, which sets forth a triple bottom line commitment to people, planet and prosperity.[140] In their 2014 Real Economy report, the GABV assessed the performance of their banks over 10 years since 2003 and demonstrated how these figures are destroying the myth that concerns about sustainability and resiliency as well as social empowerment will result in lower

[138] http://www.naturalcapitaldeclaration.org

[139] Piketty (2015).

[140] http://www.gabv.org

returns. In fact, they demonstrate higher levels of growth in loans and deposits than do traditional banks.[141]

3.13.3 *Impact Investing*

A markedly progressive solution for investors who wish to integrate their values with their investment decisions emerged in 1985, when the VanCity Credit Union in Canada responded to investors' demands for more sustainable investment opportunities, introducing the first ethical mutual fund. This fund added ethical, social, and environmental criteria to its rating benchmarks. This action marked an important transition from traditional investing, with its profit-only orientation, to *impact investing*, which also includes people, planet and profit in its success metrics. Impact investing was born, although it was only baptized with the term in 2007 during the Bellagio Summit convened by the Rockefeller Foundation.[142]

An important aspect of impact investing that could ensure its speedy adoption and evolution across all asset classes is the investors' commitment to measure and report the impact (people, planet and prosperity), and also to ensure transparency and accountability. Figure 3.14 positions *impact investing* between traditional investing and philanthropy and includes the investment criteria, metrics and risk-return considerations of its sister approaches, known by many names, which include sustainable investing, socially responsible investing (SRI), sustainable and responsible investing, programme-related investing (PRI) and mission-related investments (MRI).

Impact investing has been driven by and is very popular among individual investors and family businesses due to its progressive mind-set and freedom from regulatory obligations. However, institutional investors have also embraced and facilitated its success.[143] Most market participants, unless they are privately held, are bound by fiduciary responsibilities, and must deliver market rate financial returns, and only financial returns. Therefore, the financial performance of impact investing is still key to its wide adoption. The good news is that investments using some kind of Environmental, Social and Governance (ESG) criteria totalled US$ 21.4 trillion worldwide in 2014.[144]

---

[141] GABV (2014).

[142] https://thegiin.org/impact-investing/need-to-know/#s2

[143] Extel/UKSIF SRI & Sustainability Survey 2015 https://www.extelsurveys.com/Panel_Pages/PanelPagesBriefings.aspx?FileName=Extel-UKSIF_SRI_Report_2015

[144] 2014 Global Sustainable Investment Review by the Global Sustainable Investment Alliance. Viewed on May 14th, 2016 at http://www.gsi-alliance.org/wp-content/uploads/2015/02/GSIA_Review_download.pdf

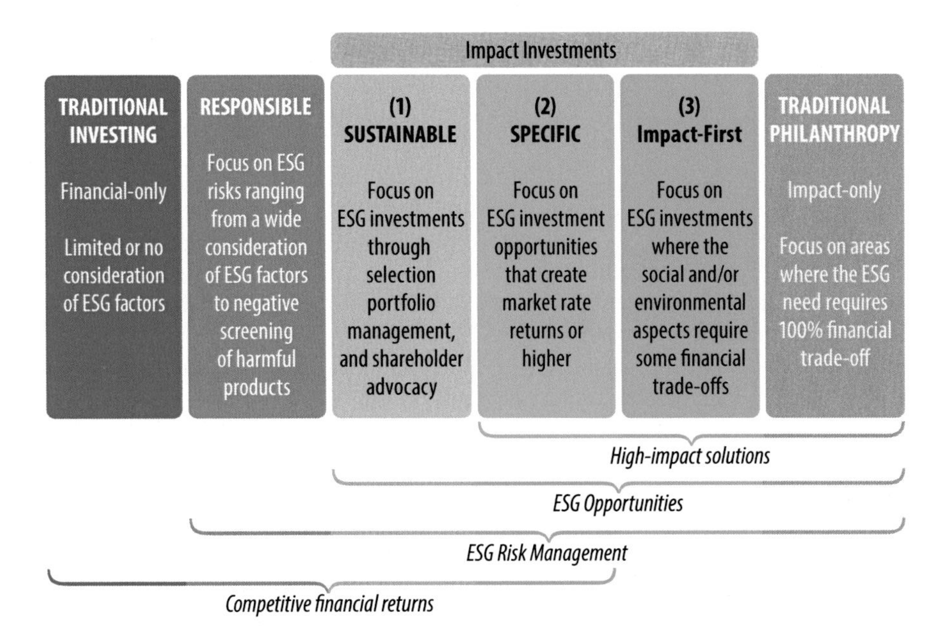

Fig. 3.14 Positioning traditional investing, responsible investing, impact investing and philanthropy (Source: Lisa Brandstetter und Othmar M. Lehner, Impact Investment Portfolios Including Social Risks and Returns [2014]. Fig. 4; based on: Susannah Nicklin, The Power of Advice in the UK Sustainable and Impact Investment Market, Bridges Ventures [2015])

3.13.4 Becoming Mainstream Is Key

In order for this industry to grow dramatically, it must become mainstream through better risk mitigation and a higher integration of more easily measurable criteria. This will enable mainstream investors, such as large institutional investors currently managing more than $20 trillion of global assets,[145] to participate. Recognition by this category of investors would help legitimize impact investing for other financial institutions, intermediaries and also policymakers.

All players, including governments, intermediaries, progressive investors and businesses, seem to be moving in the right direction. In June 2013, the G8/G20 Social (and environmental) Impact Investment Task Force (now superseded by the Global Social Impact Investment Steering Group (GSG)), was initiated by David Cameron, then UK Prime Minister, and placed under the leadership of Sir Ronald Cohen. GSG has the potential to transform our ability to build a better society for

[145] Bryce, J., & Drexler, M., & Noble, A. 2013. From the Margins to the Mainstream Assessment of the Impact Investment Sector and Opportunities to Engage Mainstream Investors. A report by the World Economic Forum Investors Industries, in collaboration with Deloitte Touche Tohmatsu. Viewed May 16, 2016 from http://www3.weforum.org/docs/WEF_II_FromMarginsMainstream_Report_2013.pdf

all.[146] This initiative has already had a significant impact on government-driven initiatives and regulations worldwide.

In October 2015, a momentous leap forward occurred when Thomas Perez, then secretary of the US Department of Labor, 'repealed restrictive guidance that prevented pension funds from engaging in impact investing'.[147] Progress has been made on the philanthropic regulations side as well. Since September 18, 2015, the US Internal Revenue Service (IRS) issued a new regulation,[148] under which foundations are allowed to invest their trust assets in mission-driven organizations aligned with the foundation's purpose, without fearing penalties for accepting lesser financial returns. However, the impact investment community is now looking with some concern at policy changes resulting from recent election changes in Britain and the United States.

3.13.5 Green Bonds, Crowdfunding and Fintech

Benign investment is not only a topic for institutional and other professional investors. Another idea is Green Bonds, intended to fuel the low-carbon transition. These bonds are a financial instrument which can be used to finance green projects delivering environmental benefits. Annual issuance rose from just US$ 3 billion in 2011 to US$ 95 billion in 2016.[149] In 2016, Apple issued a US$ 1.5 billion Green Bond, to help fund the use of green materials and more energy efficiency, such as renewable energy for data centres, becoming the first technology company to issue such a bond. The year 2016 also saw the first municipal Green Bond issuance in Latin America (Mexico City), which raised US$ 50 million to pay for energy-efficient lighting, transit upgrades and water infrastructure. In 2017, the French government announced the largest sovereign Green Bond issuance to date, EUR 7 billion, to fund the energy transition.

The 2017 OECD report suggests that by 2035 Green Bonds will have the potential to scale up to US$ 4.7–5.6 trillion in outstanding securities and US$ 620–720 billion in annual issuance for at least three key sectors in the EU, the United States, China and Japan.

This is all mostly conventional investment. However, sometimes it's outsiders and marginal players who come up with great solutions to pressing problems. But such people have difficulties finding business, state agencies or professional investors willing to finance the critical early steps for realizing their ideas.

'Crowdfunding', using electronic media to fuel investment, is a fashionable and encouraging way out of this problem. It emerged in the late 1990s, starting mostly

[146] http://www.socialimpactinvestment.org

[147] Fitzpatrick (2016).

[148] https://www.missioninvestors.org/news/irs-issues-notice-clarifying-treatment-of-mission-related-investments-by-private-foundations

[149] Hideki Takada and Rob Youngman. 2017. Can green bonds fuel the low-carbon transition? See also OECD (2017).

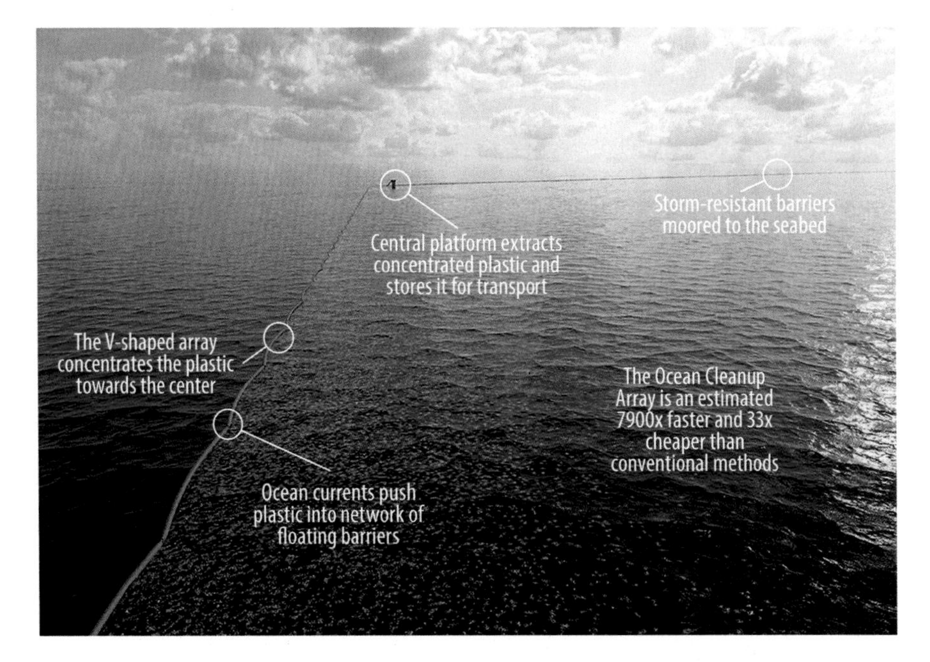

Fig. 3.15 Idea of the Ocean Cleanup Project (Source: www.theoceancleanup.com; photo: Erwin Zwart/The Ocean Cleanup)

with music and other arts, and has evolved into a whole new universe of funding opportunities for individuals and organizations. One environmental example may serve to illustrate the idea of how it can benefit the earth.

Decades of use of plastics has resulted in catastrophic pollution of our oceans. Millions of tonnes of plastic waste accumulated and are mostly concentrated in five large rotating currents, called gyres.

Off the Californian coast, there is now roughly six times more plastic than zooplankton – by dry weight.[150] People around the world began complaining, but solutions were not found. And then came an 18-year-old Dutch youngster, Boyan Slat, who invented a device for concentrating and catching plastic debris driven by ocean currents. Figure 3.15 shows the dimensions of the scheme. Slat used crowdfunding to raise more than $2.2 million to finance feasibility and prototypes to reduce plastic pollution in our oceans. Sharing his concept publicly meant his designs were challenged by others, resulting in the technical process being improved and tested in ever-larger demonstrations. The first functional prototype is under construction off coast at the Tsushima Strait of East Japan.

A different approach to linking financing with technology is called Fintech. It is designed to make financial services more efficient and highly reliable, but it also introduces a solution that may become an incremental, radical and disruptive innovation. How far it goes depends on further development of its applications, processes, products and business models within the financial services industry. The

[150] Moore et al. (2001).

core service of banking is to provide trust among participants, along with trading mechanisms that provide various financial services.

Previous technological advancements lowered the entry barrier for financial business, increasing the number of participants for both experienced and inexperienced individuals. The financial system is today being affected by 'blockchain'[151] technology, which has emerged from the philosophy of decentralization, and has the potential to shake the current banking system to its foundations. Both, venture capital spent on Fintech start-ups as well as first experiments with emerging technologies like blockchain or Ethereum[152] provide evidence that a revolution is underway. In early 2016, 'blockchain consortium R3 CEV' announced its first distributed ledger experiment using Ethereum and Microsoft Azure's blockchain as a service, involving eleven of its member banks.[153]

The underlying philosophy is the creation of trust via decentralized and shared information, which would wipe out information asymmetry among business trades. Because each block secures and locks the previous block, it is not possible to alter the previous one. A chain of trust is created, reducing risk and temptations to frauds. If open and symmetric chains of trust can be implemented, as proposed with 'blockchain' technology, the game of finance could be changed.

3.14 Measuring Well-Being Rather than GDP

The shortcomings of having GDP growth as the main objective for societal development have been addressed in numerous reports. The problems are manifold. Growth of GDP is no guarantee for non-economic objectives, rather the contrary. Furthermore, in the increasingly digitized economy, growth of GDP no longer serves to indicate increases in the number of jobs.

3.14.1 Recent Work on Alternative Indicators

In recent years, much work has been done on alternative indicators to GDP – more comprehensive indicators that would consolidate economic, environmental and social elements into a common framework to show net progress (or decline). A

[151] Blockchain, developed since 2008, is a distributed database based on the bitcoin protocol that maintains a continuously growing list of data records hardened against tampering and revision, even by its operators. Applications include crowdfunding, bitcoin transactions, supply chain auditing, and the sharing economy. https://en.wikipedia.org/wiki/Block_chain_%28database%29

[152] https://en.wikipedia.org/wiki/Ethereum – Ethereum is a cryptocurrency and a blockchain platform with smart contract functionality. It provides a decentralized virtual machine. Ethereum was proposed by Vitalik Buterin in late 2013 and the network went live on 30 July 2015.

[153] http://www.ibtimes.co.uk/r3-connects-11-banks-distributed-ledger-using-ethereum-microsoft-azure-1539044, Ian Allison, 2016-01-20.

number of researchers have proposed alternatives to GDP that make one or more of these adjustments with varying components and metrics. Others have also noted the dangers of relying on a single indicator and have proposed a 'dashboard' approach with multiple indicators. Ida Kubiszewski[154] has described many of them, including the Genuine Progress Indicator, Ecological Footprint, Biocapacity, Gini coefficient and Life Satisfaction.

The various proposed alternatives can be divided into three broad groups:

- Measures that modify economic accounts to address equity and non-market environmental and social costs and benefits
- Measures of 'subjective' indicators based on survey results
- Measures that use a number of 'objective' indicators

One such indicator, located in the first broad group, is the Genuine Progress Indicator (GPI), a version of the Index of Sustainable Economic Welfare (ISEW) first proposed in 1989. GPI starts with personal consumption expenditures (a major component of GDP) but adjusts it using about 25 different components, including income distribution, loss of leisure time, costs of family breakdown, unemployment and other negative outcomes like crime and pollution; depletion of natural resources; as well as the numerous environmental costs of GDP growth, such as loss of wetlands, farmlands, forests and ozone, and long-term damage such as climate change.[155]

GPI also adds positive components left out of GDP, including the benefits of volunteering and household work. By separating activities that diminish welfare from those that enhance it, GPI better approximates sustainable economic welfare. However, GPI is not meant to be an indicator of sustainability. It is a measure of economic welfare that needs to be viewed alongside biophysical and other indicators. In the end, since one only knows if a system is sustainable after the fact, there can be no direct indicators of sustainability, only predictors.

If the same method of input/output tables were to be used to calculate GDP, the entire process would have to be adjusted. The tables would have to distinguish between the economic activities that add to human well-being versus those that subtract from it (see Fig. 3.16). Another major change would have to be the inclusion of goods and services that are not within the economic market but do have a large influence on human well-being. Over the past few years, various groups, including the United Nations and the World Bank, have been working on creating national accounts that incorporate ecosystem services. Some of these efforts modify the input/output model to incorporate services provided by nature.

Over the past few decades, ISEW or GPI have been calculated in around 20 countries worldwide. These studies have indicated that in many countries, beyond a certain point, GDP growth no longer correlates with increased economic welfare. The trend is similar in many countries. GPI tracks GDP pretty closely as a country

---

[154] Kubiszewski (2014).

[155] Components of GPI, genuineprogress website (EU): https://genuineprogress.wordpress.com/the-components-of-gpi/

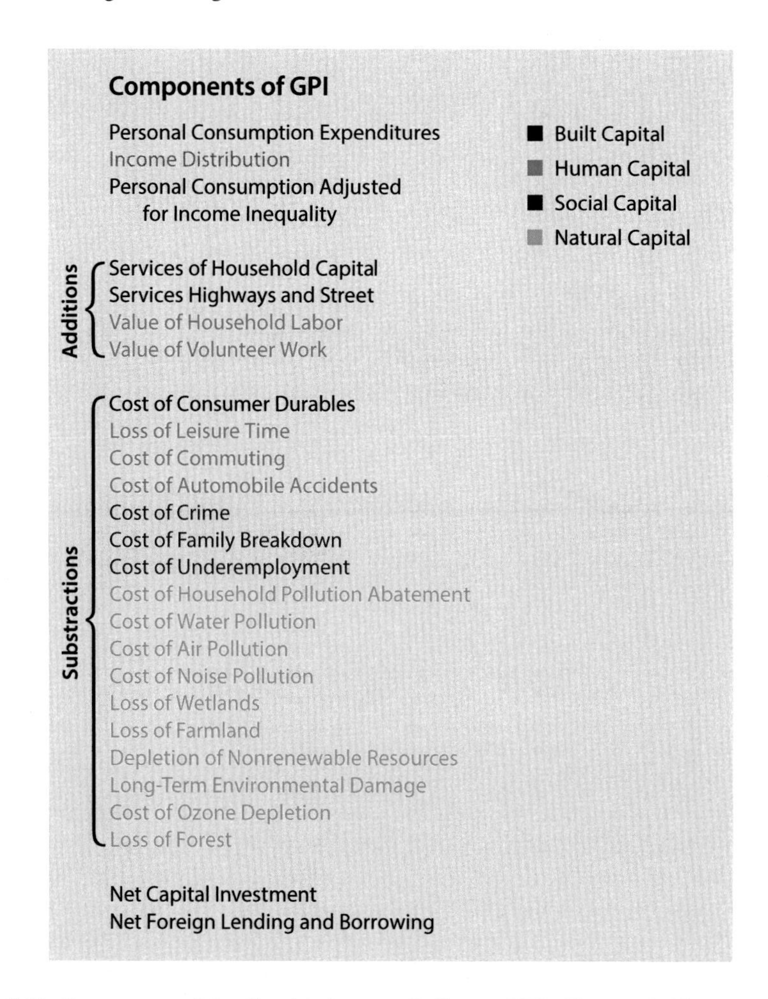

Fig. 3.16 Components of the Genuine Progress Indicator (GPI) (Source: Kubiszewski et al. 2013)

develops, but at a certain point the two diverge. In the United States it happened in the mid-1970s, while in China in the mid-1990s. GDP keeps growing while GPI levels off or decreases.

Recently, a global GPI was also estimated using GPI and ISEW data from 17 countries, containing approximately 53% of the world's population and 59% of the global GDP. On the global level GPI/capita peaked in 1978 (Fig. 3.17). Interestingly, 1978 is also around the time that the human ecological footprint, a biophysical indicator that measures humanity's demand on nature, exceeded the earth's capacity to support humanity. Other global indicators, such as surveys of life satisfaction from around the world, also began to level off around this time. In fact, a strikingly consistent global trend suggests that as income increases, well-being often decreases,

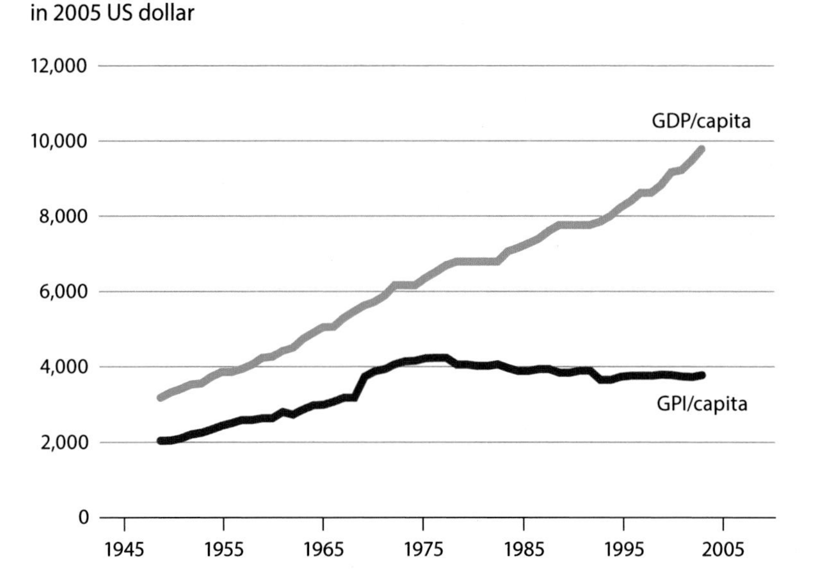

Fig. 3.17 Global GPI/capita and GDP/capita. GPI/capita was estimated by aggregating data for the 17 countries for which GPI or ISEW had been estimated, and adjusting for discrepancies caused by incomplete coverage by comparison with global GDP/capita data for all countries. All estimates are in 2005 US$ (Source: Kubiszewski et al. 2013)

accompanied by rising rates of alcoholism, suicide, depression, poor health, crime, divorce and other social pathologies.

An important function of GPI is to send up a red flag at that point. Since it is made up of many benefit and cost components, it also allows for the identification of which factors increase or decrease economic welfare. Other indicators are better guides of specific aspects. For example, Life Satisfaction, as determined by surveys, is a better measure of overall self-reported well-being. By observing the change in individual benefit and cost components, GPI reveals which factors cause economic welfare to rise or fall even if it does not always indicate what the driving forces are behind this. It can account for the underlying patterns of resource consumption, for example, but may not pick up the self-reinforcing evolution of markets or political power that drive change.

Recently, two state governments in the United States have adopted GPI as an official indicator, the states of Maryland and Vermont. In addition, the data necessary to estimate GPI is becoming more available in many countries and regions. For example, remote sensing data allow better estimates of changes in natural capital and surveys of individuals about their time use and life satisfaction are also becoming more routine. New means of measuring inequality are being developed, and more detailed data are being collected on the costs of crime, family breakdown, underemployment and other measures that might be used in GPI in the future. The bottom line is that the costs of estimating GPI are not particularly high, the data limitations can be overcome, and it can be relatively easily estimated in most countries.

3.14.2 Divergence Between GDP and GPI

GDP was created in the United States during the 1930s and continued in use after World War II, when the world needed to repair its built infrastructure and financial systems. Natural resources were perceived as abundant; inadequate access to infrastructure and consumer goods represented the main limit to perceived improvements to human well-being. During this time, it made sense to create an indicator that ignored relatively abundant natural resources and the distribution of wealth, to focus solely on increasing the production and consumption of market goods and services, which were relatively scarce.

However, as a result of our success, the world has changed dramatically over the past few decades. We now live in a world full of human infrastructure. The human footprint has grown so large that, in many cases, limits on the availability of natural resources now constrain real progress more than limits to consumer goods.

Between roughly 1950 and 1975, GPI per person for the majority of countries was increasing. Much of this was due to the rebuilding effort after World War II when consumption and built capital were the limiting factors for improving well-being in many countries and environmental externalities had not yet become significant. By the mid to late 1970s, much of the infrastructure had been rebuilt. However, rising income inequality and increasing external environmental costs began to cancel out the growth in consumption-related benefits, causing GPI/capita to level off.

GPI is not a perfect measure of overall human well-being since it emphasizes economic welfare and leaves out other important aspects of well-being. It is, however, a far better indicator than GDP, which was not designed to measure welfare at all. Societal well-being or welfare ultimately depends on stocks of natural, human, built and social capital, and because the GPI makes additions and deductions to GDP to reflect net contributions to these stocks, it is a far superior measure of economic welfare than GDP. The disconnect between GPI and GDP, beginning in 1978, shows the aspects of our well-being that have been declining since that time. It also provides focus areas where societal improvement is necessary and possible.

3.14.3 Towards a Hybrid Approach

All the approaches mentioned above have positive and negative aspects. So the question becomes: can we construct a hybrid indicator that incorporates most of the positive aspects and minimizes the negative? As Costanza et al.[156] conclude, 'The

[156] Costanza et al. (2014a).

successor to GDP should be a new set of metrics that integrates current knowledge of how ecology, economics, psychology and sociology collectively contribute to establishing and measuring sustainable wellbeing. The new metrics must garner broad support from stakeholders in the coming conclaves'.

Against this backdrop, one potential hybrid, the Sustainable Well-being Index (SWI), could be a combination of three basic parts, each covering the contributions to sustainable well-being from the dimensions of economy, society and nature.[157]

Net Economic Contribution The GPI can be thought of as a measure of the net contribution of economic (production and consumption) elements to well-being. It weighs personal consumption by income distribution, adds some positive economic elements left out of GDP, and subtracts a range of costs that should not be counted as benefits. Although some costs to natural and social capital are included in GPI, many others are missing (e.g. loss of community cohesion due to the social disruptions caused by economic growth). Conversely, we need a way to measure and include the positive benefits to well-being from natural and social capital. The current GPI is in need of being supplemented with additional cost estimates from the SDGs, including its targets and proposed indicators, as well as measurements of the positive contributions of natural and social capital.

Natural Capital/Ecosystem Services Contribution The positive contributions of natural capital and the ecosystem services it provides have been estimated in spatially explicit form and can be valued in different units, including monetary units.[158] These can be estimated at the country level, as well as at subnational and regional scales. For example, the Wealth Accounting and Valuation of Ecosystem Services (WAVES) project of the World Bank[159] is actively pursuing this agenda, as are other initiatives, including the new Intergovernmental Science-Policy Platform on Biodiversity and Ecosystem Services, The Economics of Ecosystems and Biodiversity (TEEB) and the Ecosystem Services Partnership.[160]

Social Capital/Community Contribution The positive contributions to well-being from social capital could be captured via surveys of the various components of life satisfaction. For example, the World Values Survey as well as regional barometers (e.g. Eurobarometer, Afrobarometer, etc.) ask questions about trust and other aspects of social capital. However, we may need to add additional survey questions

[157] Costanza et al. (2016). For further considerations see also Fioramonti (2017).

[158] Costanza et al. (1997, 2014b).

[159] Wealth Accounting and Valuation of Ecosystem Services (WAVES): https://www.wavespartnership.org/

[160] Intergovernmental Science-Policy Platform on Biodiversity and Ecosystem Services (IPBES)2014: http://www.ipbes. net/; The Economics of Ecosystems and Biodiversity (TEEB): – http://www.teebweb.org); and the Ecosystem Services Partnership (ESP): http://www.fsd.nl/esp

that ask explicitly about the value of community and social capital, in addition to individual life satisfaction.

3.15 Civil Society, Social Capital and Collective Leadership

In Chap. 1 of this book (Sect. 1.10), the UN 2030 Agenda for Sustainable Development was outlined. It chiefly consists of the 17 Sustainable Development Goals (SDGs). In the real world of business and associated politics, priority most certainly will be given to the economic and social goals, thus jeopardizing the healthy stabilization of the climate, of the oceans and of biodiversity (SDGs 13–15). To balance this bias, using the language of the 2030 Agenda, the SDGs are highly interconnected and should be seen as a whole.

Whereas governments and businesses have an agenda of their own and all too often cater to vested interests, civil society can play an important role in pushing towards a sustainability transformation. On the other hand though, this cannot be civil society alone. These challenges are complex, systemic and broad in scope; so it is only through collaboration of all sectors that change can be driven successfully. Mary Kaldor[161] defines civil society as 'the process through which individuals negotiate, argue, struggle against or agree with each other and with the centres of political and economic authority'. She outlines how the roles and meaning of civil society has changed over time – starting with movements in South America and Eastern Europe opposing militarized regimes, and listing the definitions, most commonly used at present to describe global civil society. But one can just as well go deeper into history and broaden these definitions to include trade unions, abolitionists, suffragettes and many others. Such a definition of civil society organizations (CSOs) as inclusive of social movements concerned with more just and equal power distributions, is more historically correct.[162]

On the other hand, trust, solidarity, collaboration and sustainability thoughts are not necessarily dominant in the wide spectrum of CSOs. Attention should also be paid to undesired outcomes of successful citizens' movements. During the first year or so of the 'Arab Spring', commentators around the world celebrated this encouraging new feature in an otherwise rather authoritarian and rigid landscape. But the enthusiasm faded when either violent Da'esh ('Islamic State') groups or new authoritarian regimes took over, civil wars raged, and the Near East turned into the number one problem area of the world.

One must also recognize that populist movements including radical right-wing ones have adopted the communication techniques of civil society. The violent movements have developed skills taking advantage of those media for aggressive propa-

[161] Kaldor (2003); here: 585.

[162] For an historic overview see Tilly (2004); for the global context see Keane (2003)

ganda. Remember that it is angry emotions that are most likely to get viral in the social media.[163]

In times where 'alternative facts' have become a concept of argumentation for the angry ones, there is an urgent need for a counterbalance from what Mary Kaldor calls 'political bargaining' with a public and 'good-tempered' conversation. This conversation requires reason and sensitivity and not just conflicting interests and passions. It provides a solid basis for building up the *social capital* which, according to Francis Fukuyama, exists when the abstract idea of 'relationship' is replaced by an actual, collaborative and communicative relationship between two or more human beings.[164] This creates Social Capital, which in turn gives rise to concepts such as trust, networks and civil society. It is this openness to uncertainty and the ability to keep meeting each other collaboratively which holds the best possibility of bringing about large systems changes.

3.15.1 Public Conversation: The Concept of Citizens' Assemblies

Establishing a 'good-tempered' conversation in order to engage citizens in a public debate is a first and essential step towards a new understanding of reintegrating citizens into the public sphere. Modern democracies have developed into elitist systems that have most recently provoked strong counter-movements by those feeling left out. The Brexit and the Trump election are two of the most striking examples. However, a very common mistake here is to confuse the need for a public debate with demands for direct decision making – the latter all too often leads to uninformed voting practices and lacks an informative and transparent debate altogether. Referenda (or using elections the same way) often lead to perverse decisions that are neither for the good of society as a whole nor for those who voted in favour of them. Therefore, a *real* public conversation is needed, in which people feel included and represented – but most of all *informed*. An impressive example for an institutional discussion allowing for such a conversation is Ireland's Citizens' Assembly, introduced in 2012. Citizens are randomly selected to participate in order to debate topics and give an informed recommendation to their parliament for decision making.

Politicians can learn their citizens' needs and desires, their fears and wishes. Citizens, in turn, are neither left out of the process, nor are they suddenly thrown into an unfamiliar arena, asked to vote on something on which they have never had a chance to properly deliberate. In the case of the Citizens' Assembly, participants can form their opinions as they learn, discuss and exchange arguments. Ireland's case is built on the same logic as Ned Crosb's *Citizen's Jury* in the United States and

[163] Fan et al. (2014).

[164] Fukuyama (2001).

Peter Dienel's *Planungszelle*,[165] both concepts developed in the 1970s. They are based on the claimed need to engage citizens in decision making and planning processes – and they all do so by lot selection, as with a jury. This brings such processes close to the often-cited origins of democracy in ancient Greece, where politicians also were assigned by lot, not by vote. This fact has been absent in the evolutionary process of modern democracies. This important difference has partially allowed for the political arena to develop into something so disconnected from society. Today, apparently democratic methodologies, like referenda, are too easily used by populist movements to make false promises that such rapid, undiscussed actions will give 'power to the people'.

3.15.2 Building Up Social Capital: Multi-stakeholder Collaboration

An informed public debate provides a good basis for an active civil society; but it is not sufficient to tackle the world's current challenges in all their complexity.

When it comes to shifting the currently dysfunctional world direction into a more functional one, no single actor – neither civil society nor politics nor business – will be able to deliver the entire solution.[166] Instead, each actor must contribute a different but essential piece of knowledge. At the same time, the organizational cultures of civil society, governments and business are very different, and so are their leadership cultures. It is important for all three camps to acknowledge that they operate separately from one another and that what takes place outside of their spheres of familiarity should not automatically be looked at with suspicion. It is through collaboration between these three, interlinked systems that new forms of social capital can be created. Multi-stakeholder collaboration opens a way for the innovation and collaboration needed for civil society, business and government to grow towards mutually supportive learning journeys.

A multi-stakeholder collaboration should be characterized by:

- Multiple actors, often with conflicting interests, who need to align around a joint improvement approach.
- The effectiveness of the collaboration is dependent on engaging actors, who would not normally work together, into a joint approach.
- Multidimensional problems tend to require solutions that are typically *complicated*, *complex* and even *chaotic* – owing to unforeseen market or political influences.[167]

[165] http://www.planungszelle.de

[166] Petra Kuenkel and Kristiane Schaefer: 2013. Shifting the way we co-create. How we can turn the challenges of sustainability into opportunities. The Collective Leadership Institute.

[167] Snowden and Boone (2007)

Multi-stakeholder collaboration is a systems approach, and as such can be seen as a complex yet purposeful endeavour to affect social change. It has the potential to shift or rearrange existing societal settings and overcome organizational limitations. Leadership, in this context, is a co-creative process that often begins with a small group of dedicated initiators and aims at profound collective change.

Even the greatest visions for change are futile if not enough stakeholders are prepared to commit to action. Effective multi-actor settings therefore require sufficient engagement of stakeholders – the powerful and the less powerful, the influential and the affected.

Conscious collaboration – setting up a temporary or lasting system of multi-stakeholder actors – is a way of creating life. A people-centred and planet-sensitive future requires us to build many such nested *collaborations.*

3.15.3 A Case of Collective Leadership: The Common Code of the Coffee Community

Author Petra Kuenkel[168] uses the *Common Code of the Coffee Community (4C)* as an example of a multi-stakeholder setting with a 'collective leadership' approach. Something called the *leadership compass* (Fig. 3.17) was used as a navigating tool for process planning.

The 4C developed out of a cross-sector partnership between three stakeholder groups – the coffee trade and industry, coffee producer organizations and international civil society organizations. The 4C association is a remarkable example of the creation of a global community that joined forces to improve the social, environmental and economic conditions for those who earn their living from coffee. The most important improvements were the application of a code of conduct, support mechanisms for farmers and a verification system.

The 4C initiative, like many other multi-stakeholder initiatives, moved through four different phases.[169] Even though it is important to keep the six dimensions of the *Collective Leadership Compass* in a healthy balance throughout the process, each phase requires a difference in focus (Fig. 3.18).

Phase 1 (preparing the system for collaboration) was about shaping the idea in dialogue, understanding the context, and initiating the multi-stakeholder initiative. In the 4C initiative, the emphasis was placed on building trusting relationships, testing existing and possible future cooperation. Using the compass for planning and process management helped actors from all sectors stay in dialogue around the initial idea to influence the mainstream market towards greater sustainability. Because people met repeatedly to collaborate on similar issues and specific topics with regard to coffee and sustainability, the idea of developing a mainstream standard

[168] Kuenkel (2016).

[169] Kuenkel et al. (2009).

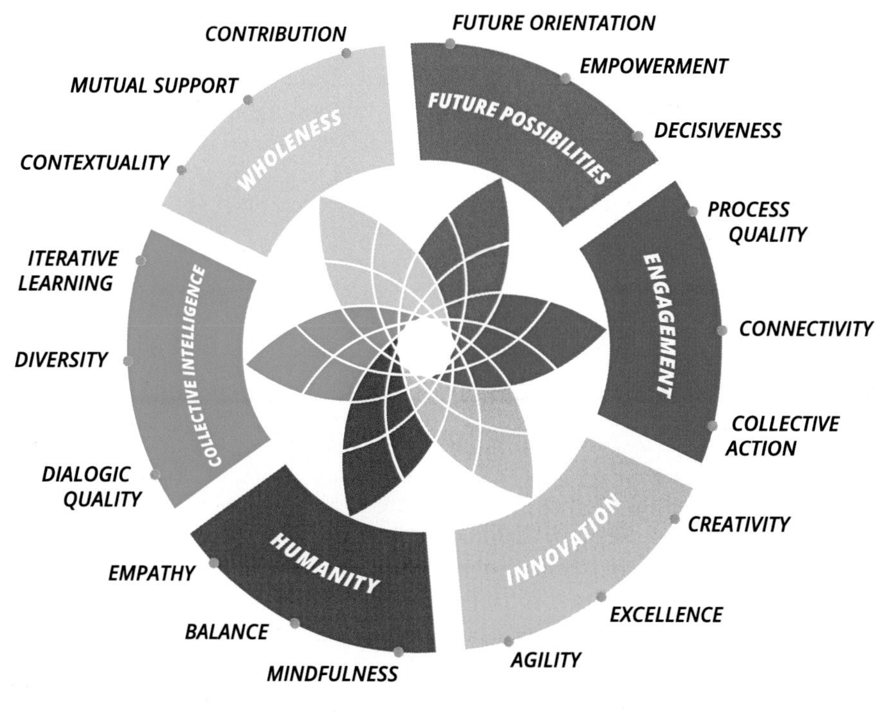

Fig. 3.18 The Collective Leadership Compass by Petra Künkel served as a guide for the multi-stakeholder collaboration process in developing the Common Code of the Coffee Community

slowly began to take root. Despite the challenges and the absence of easy answers, the initiative found support in many different countries in Asia, Africa and Latin America. People realized that there was a real chance to have a structural effect on the imbalances that are part of coffee production.

Phase 2 (building a collaboration system) was about reshaping the goal, clarifying resources, creating a structure for the initiative, and agreeing on a plan of action. Selecting this group of stakeholders was based on finding the right balance between 'engaging the interested' and 'engaging the official representatives'. The result was an agreed upon implementation plan, a budget plan for future financial contributions by the industry involved, and an allocation of roles between the stakeholders.

Phase 3 (Implementing collaboration) requires a regular reinforcement of the power of the potential impacts of stakeholder meetings, which were not free of conflicts. Mistrust never completely disappears, yet all stakeholders learned to remain collaborative and move towards tangible results.

Phase 4 (Taking collaboration to the next level) After 2 years, the standard had been developed and the initiative moved into this phase. In 2006, the stakeholders unanimously agreed to establish a non-profit organization that would become the future formal structure for the initiative, that is, a global membership organization (the 4C Association), dedicated to implementing sustainability in the coffee sector and open to coffee chain participants ranging from small coffee farmers to large roasting companies as well as to all others on a supportive base.

3.16 Global Governance

3.16.1 Introduction: The UN System and Forward-Looking Ideas

Much of the work of the Club of Rome is related to global problems and initiatives. Many of the ideas presented in Chap. 3 of this book directly or indirectly require some coordination or ruling at the global level.

In Sect. 2.5, it was said that the 'philosophy' of the nation state, emerging during the phase of the 'empty world', has to be overhauled in many respects which include some legal instruments of *global governance*. This is not new. When the United Nations were founded in 1945, everyone knew that the horrors of a World War had to be avoided in the future and that the nations of the world must come together building a transnational, a global institution with powers that in certain cases could override those of nation states. Our book is not targeted on the functioning or malfunctioning of the United Nations system. But one would surely agree that despite all its shortcomings, the UN is a *must* and a blessing.

What has to be targeted, however, are ideas and institutions helping the global coordination of policies supportive of the 17 Sustainable Development Goals. For that to happen, options both below and beyond the UN system have to be considered.

At the outset, two different approaches to global governance and international cooperation may be mentioned. One stems from the World Future Council (WFC), an NGO founded by Jakob von Uexkull, founder and original sponsor of the Right Livelihood Award; the other one is Paul Raskin's 'Great Transition'.

The WFC has worked on a *Global Policy Action Plan* (GPACT)[170] for some years now, with justice at its core. A 'Roadmap to a Future-Just World' was conceived, which comprises seven sections, including peace and security, equity and dignity and climate

[170] World Future Council Foundation (2014). Global Policy Action Plan. Incentives for a Sustainable Future. Braunschweig: Oeding Print. For more go to www. worldfuturecouncil.org or mail to the program's coordinator, catherine.pearce@worldfuturecouncil.org

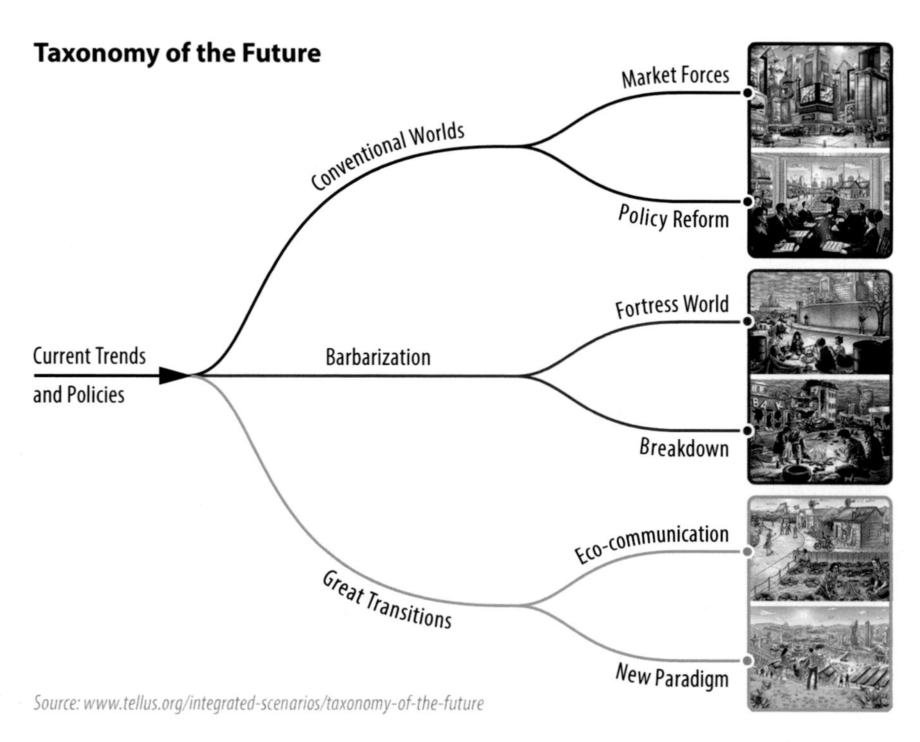

Fig. 3.19 Paul Raskins 'Taxonomy of the Future' shows two desirable, two nasty but perhaps tolerable and two terrible options for the future (Courtesy Paul Raskin, www.tellus.org/integrated-scenarios/taxonomy-of-the-future)

stability. The WFC brochure on GPACT summarizes many best policies from around the world, such as Hungary's Act establishing an ombudsperson for fundamental rights, Belo Horizonte's Food Security Programme (Brazil), or Exeter University's 'One Planet MBA'. Moreover, the programme specifies principles for *future just lawmaking* which are clearly addressed to national lawmakers around the world.

People and associations working on a benign global governance should help build more public acceptance for the GPACT philosophy. Once such movements gain influence and power, discussion will flourish about the even more ambitious task of *global governance for a sustainable world.*

A related and equally ambitious line of thinking comes from Paul Raskin, mentioned in the section 'Linking Chaps. 1 and 2' of this book. His 'Journey to Earthland' confronts 'twenty-first-century challenges hobbled by twentieth-century ideas and institutions. Zombie ideologies – territorial chauvinism, unbridled consumerism, and the illusion of endless growth – inhabit the brains of the living'.[171] Raskin depicts three major trajectories – Conventional, Barbarization and Great Transition – each opening two further options, as shown in Fig. 3.19.

[171] Raskin (2016). Quote from p. 21.

Raskin's latest booklet sketches out the ominous trends resulting from governance by markets, which will lead to 'barbarization', or at least highly unpleasant conditions. He proceeds to compare such negative trends with the Great Transition options, and names nine parameters: population, gross world product (or GWP), work time, poverty, energy, climate, food, habitats and freshwater withdrawals. For the year 2100, all nine, in terms of sustainability and happiness, are much worse in a market-driven world than in one moving into transition.

This leads to the broadening conviction that the mechanisms of pure market economies and weak interventions by well-meaning but feeble 'reform' politicians are outdated. The Great Transition, by contrast, leads to the Planetary Phase of *One World and Many Places*, with the governance 'principle of restrained pluralism'.[172] Following that principle, wasteful consumption and population density should recede, international trade and economic turnover could stabilize, while education, leisure time, spirituality and social justice get more and stronger.

Currently, the Journey to Earthland is a dream, of course, but a necessary one, if compared with the barbarization alternatives.

3.16.2 Specific Tasks

Global governance will in most cases consist of specific tasks. In the UN system, these tasks consist currently of actions such as local military interventions with the purpose of peace keeping; trade rule-setting by the WTO; loan and aid programmes by the UNDP and the World Bank; or campaigns to combat infectious diseases orchestrated by the WHO. All such actions find broad support by the nation states, the civil society and the business community.

But today of course there are new challenges, as outlined in this book. Section 1.10 pointed to the UN 2030 Agenda with its 17 Sustainable Development Goals, and Sects. 1.5 and 3.7 addressed global warming and its remedies. In Sect. 1.6.1, 'technological wild cards' were mentioned, notably synthetic biology, geoengineering and artificial intelligence, all of which have the potential to surpass human capacities and to get out of control. This list of three were selected by Cambridge University's Centre for the Study of Existential Risks, but are far from exhaustive. It appears unavoidable that the international community should set up a globally mandated institution or network for technology assessment. Contrary to fashionable beliefs within the 'innovation community', this will not be intended as an instrument to block or bureaucratize technological progress, but rather will offer some early warnings and suggest some redirection of the definition of that progress. In financial terms, this means avoiding huge misallocation (and later destruction) of capital.

A different task, already alluded to in Sect. 3.11, is designing and implementing mechanisms to rebalance public and private goods. That balance was lost during the

[172] Ibid., pp. 84–87.

1980s and 1990s, when markets were freed from many legal restrictions and became truly global, while the law essentially remained national – creating a massive imbalance to the favour of markets and disfavour of the law.

The term 'Markets' applies to two different animals. There are the markets of goods and services, and their success is mostly determined by quality and prices. This is generally a benign mechanism that usually leads to a steady increase in quality and affordability. This kind of market can be global, perhaps with exceptions for 'infant industries' needing some early protection and for processes with high environmental impacts needing some cautious restrictions.

The other type of markets are *financial* markets. They have become extremely powerful and tend to contain strong speculative features. Estimates exist (see Sect. 1.1.2) that out of every 100 dollars travelling across borders (essentially at the speed of light), only two actually pay for goods and services. The dominance of financial markets forces lawmakers in all countries to set rules that allow or support highest possible returns on investments (RoI). In practical terms, this means reducing the tax burden for business, reducing regulations, keeping prices low for infrastructure use, for land, energy, water and other resources, even subsidizing investors' activities. Small wonder that this worldwide trend often works to the detriment of public goods such as natural resources or public infrastructure, for which the state would normally be responsible. Here the global governance challenge is simply to return to a better balance between public and private goods and welfare.

Social equity can also be seen as a public good and is also suffering. We observe in the process of globalized capital markets that direct taxation, chiefly on capital, is receding, while indirect taxes are rising, such as the value-added tax (VAT) which hits poor families (who cannot evade to tax havens) much more than it impacts rich ones and corporations. Again, global governance should reduce the role of tax havens and should aim at harmonizing tax rates on profits, capital and financial flows.

3.16.3 COHAB: Cohabitation Mode of Nation States

A top-down reform of the United Nations with strong rules of global governance is unlikely to happen. So it may be fruitful to consider different approaches. One is the idea of 'cohabitation' among the nearly 200 states of the world. Gerhard Knies, a scientist from Hamburg, Germany, and one of the initiators of the Club of Rome's 'Desertec'[173] idea, and now organizer of the ViableWorld Design Network, has engaged in an operational strategy meant to lead to a viable world.[174] It consists of a tolerance-based modification of the architecture of the United Nations with the imperative goal of protecting and developing *the global commons*, for example,

[173] More at http://www.desertec.org

[174] Gerhard Knies. 2016. Model of a Viable World for 11 Billion Humans and Future Generations. Typoscript.

ensuring a more stable climate, along with the ot*her environmental* and developmental goals of the 2030 Agenda.

At a time where an electoral majority in Britain finds the EU too large a body for its purposes and votes to leave so as to 'take control' of national affairs, any global governance idea may sound impossible. But the global challenges we are facing don't disappear just because some voters don't recognize them. Supra-national cooperation simply has to take place, and with increasing intensity and scope.

Knies calls his approach the *Cohabitation Model for a Viable World.* Cohabitation means that nations and other geographical units will voluntarily organize ways of living together, instead of combating or ignoring each other.

A full world implies certain limitations on the sovereignty of nation states. The UN comprises nearly 200 nation states. Each state's internal sovereignty does in fact limit, to some extent, the external sovereign rights of all the other states. In the *full world,* this has become a serious problem. Moreover, any state's *internal* sovereignty is affected by the ecological footprint of each individual human being. Each kilogram of CO_2 emitted from any of the 7.5 billion persons on the planet affects everyone else on Earth, including all future generations.

Cohabitation means to make the best of it, and to optimize the connection. From the traditional rivalry attitude between nations, human society must rise to the vision of a global community. How would that work? Knies suggests five consecutive stages:

Stage 1: Intergovernmental Panel on Habitability of Planet Earth – IPHE
A first step would be the upgrading of the Intergovernmental Panel on Climate Change (IPCC) into an Intergovernmental Panel on the Habitability of Earth[175] (IPHE), to prepare the basic information required for a global contract on restoring and maintaining sufficient habitability of our planet. Accession to the IPHE would be voluntary, but some incentives can be created to rewarding joining up.

Stage 2: National Ministries for Global Cohabitation
As a second step, Knies proposes that each state create national ministries for global cohabitation. Their tasks would be to identify critical developments around the world, and to bring them to the attention of their national governments in order to address them through national policies.

Stage 3: International Cohabitation Conferences – Like Climate Conferences
Cohabitation ministers of a few cohabitation-ready nations can begin to explore how to unify various ideas, policies and national capacities so as to build a viable world. The ministers could organize 'cohabitation conferences' in order to brainstorm on how to address a broad scope of interconnected issues like climate, water, food, prosperity, population growth and other threats to the habitability of Earth. They could develop cohabitation rules and cohabitation goals, and gradually attract more nations.

[175] Gerhard Knies. 2017. Proposal to create an Intergovernmental Panel on Habitability of Earth for Humanity, IPHEH, www.ViableWorld.net, Typoscript.

Stage 4: Nation States Mutate from Rivals to Cohabitants
Nation states can begin to devote a growing fraction of their military budgets to projects reducing ecological damage and advancing human development at home and abroad. Their military forces could gradually be replaced by the people and infrastructures necessary to safeguard a viable world.

Stage 5: Cohabitation Based Global Governance
Expanding the cohabitation model for nation states will increase humanity's ability to solve our pressing global problems, whether new or already recognized, in an organized and constructive way. More and more nations will join this 'Viable World Alliance', which will see humanity as one entity, rather than as currently, a set of 200 independent and often rival national groups. Staying outside the Alliance would become embarrassing, especially if the Alliance successfully manages to build a viable world while reducing the military sector, and using its expertise at social unity to increase viable social as well as ecological structures on this planet.

The COHAB model is so far clearly a dream for a global political innovation. Still, it has a goal: to one day supersede the United Nations system, and perhaps will be called United Humanity. It may maintain many of the Specialized UN Agencies, but it would have to empower them with sanction mechanisms wherever global issues are involved. This is the essential meaning of global governance.

For all of humanity to have a survival plan does not require a global *government;* just some parameters for how *governance* would work. Of course, certain rules and codes for global cohabitation will be needed. Since conditions have changed in the Anthropocene, these codes would supersede the UN Charter. Democratic processes should be maintained and strengthened, but under the principle of *subsidiarity*. That is, matters that affect the local level should be organized and regulated there. Higher levels such as provinces, countries and geographical regions should have their respective democratic representation. But global issues should be decided at the level of the above mentioned Alliance, always respecting, of course, the needs and priorities of people at the regional, national or lower levels.

3.17 National-Level Action: China and Bhutan

Sustainability policies are foremost national policies. Of course, the success stories told at the outset of Chap. 3 are subnational. Other chapters are offering solutions for the business world or for the international level. Regarding the national level, it is impossible to go through almost 200 countries of the world. Instead two countries will stand as examples. They can be seen as extremes regarding population size and density, industrialization and importance to the world trades: China, the giant, and Bhutan, the dwarf. Both, in their specific ways, have shown remarkable strategies of handling the challenges of sustainable development. China has chosen a strategy of rapid industrialization and economic growth, and lately, of 'greening' its economy.

Bhutan has chosen a radical environmental protection agenda while proclaiming its people's happiness as more important than economic turnover.

3.17.1 China and Its 13th Five-Year Plan

China is undergoing profound changes. The period of heavy industry, cheap mass manufacturing and aggressive exporting is flattening out. At the same time, massive pollution of air and water is plaguing its people, and the demand for quality food is exceeding supplies. Earlier expectations of continued double-digit growth rates have collapsed under these new realities, and investors and speculators lost a lot of money. This, in a nutshell, was the situation in 2015, the year China released its 13th Five-Year Plan.

Five-year plans are designed by top levels of the State and the Party, which decide on regulatory directives for implementation at provincial and local levels. Since 2006, the name of the 5-year plans has been changed to 'Guideline', in order to indicate that the people's will and the markets will also influence the development during the coming 5 years. However, for an international readership we continue to use the more familiar term of 5-year plans. Starting with the 11th Five-Year Plan (from 2006 through 2010), these templates have included a strong emphasis on local environmental improvement. The 12th Five-Year Plan, from 2011 through 2015, added a strong component of decarbonization.

The 13th Five-Year Plan, adopted in 2015, has greatly enhanced the need for reducing the carbon dependency of China by setting ambitious targets for renewable energy, along with further improvement of energy efficiency. This is in line with China's commitment to the 2015 Paris Agreement on climate protection. Resource efficiency, in the framework of a circular economy, is also emphasized.

Moreover, several regional strategies have been included such as the importance of natural eco-systems in the further development of the Beijing-Tianjin-Hebei region, the Yangtse Delta and the Zhujiang Delta. Following to a large extent UNEP's 'Green Economy' Programme, the 13th Five-Year Plan formulates the concept of 'Eco-Civilization' in the development of industrial zones and city-township clusters.

For industry, the Plan introduces the concept of *green manufacturing*, laid down the 10 years' vision *Manufacturing Action Guide of 'Made in China 2025'*.[176]

Introducing Eco-Civilization has no prescribed methods so far. A good start is to assess the current status of the environment and defining its ecological limits ('control lines'), especially its base (red) line as scientifically and rationally as possible. Moreover, measures are sought that help restore the ecological quality. All these needs to be discussed with the local residents or farmers, and proper training and expertise should be offered.

[176] http://news.china.com/domestic/945/20150519/19710486.html

Control lines should be scientifically drawn at district levels. Taking economic growth into account, local authorities might, for example, seek to offset the consumption of green land against green land regained through restoration measures, such as man-made parks and reclaimed or afforested hills, although of course there are important sustainability differences between, for example, a man-made forest and a natural one.

In 2015, China selected four cities to establish 'Natural Capital Balance Sheets'[177] as trial cases. There, the 'System of Environmental-Economic Accounting 2012' from UNEP is referenced, which requires accurate data. This task requires overcoming the barriers that made sharing data among varied governmental departments difficult. In other words, the programme of *Eco-Civilization* carries the ambition of moving from slogans to quantitative and measurable actions.

Farmers are receiving explicit benefits, namely, to use allocated land for owning houses and to transfer and trade land. This option opens a fast track to gaining prosperity, which was not the case for farmers in the past.

A new social challenge is *Internet+*,[178] allowing direct sales, thereby diminishing income opportunities of traditional trades. Clearly, the Internet has given rise to e-commerce by new giants such as Alibaba, Taobao, Jingdong and Alipay.com. Their B2B, C2C and e-pay portals have successfully connected manufacturers, consumers and banks directly but have created steeply rising demand for new transport and logistics infrastructures, eating into the remaining treasures of natural landscape.

China is now experiencing the Western trend of establishing food supply chains using refrigeration and centralization. It is also jumping into new opportunities for municipal farming such as vertical farming, hydroponics, aeroponics and Community Supported Agriculture (CSA) close to urban consumers. Food safety too has a high profile in the 13th Five-Year Plan.

In the manufacturing sector, China is currently troubled with major overcapacities in heavy industry and labour costs approaching those of the OECD countries. The response, indicated in the 13th Five-Year Plan, is called 'Made in China 2025', which consists of an ambitious transformation towards the big data trend currently embraced by the United States, Japan and Germany, which aims at a consistent real-time information flow between all involved parties. The Chinese concept also includes a strong emphasis on green design with high resource efficiency, entire green product life cycle management decarbonization, and a highly efficient, clean, recyclable green production system. This is partially a response to the fact that China's many cities are suffering from horrendous air pollution.[179]

---

[177] http://www.gov.cn/zhengce/content/2015-11/17/content_10313.htm; Five cities are Hulun Buir in inner Mongolia, Huzhou in Zhejiang, Loudi in Huanan, Chishui in Gui Zhou, Yanan in Shaanxi.

[178] "Internet +" is a Chinese term since 2015, combining the Internet with any traditional business turning it into a new business model.

[179] Chinese Ministry of Environment, http://www.ocn.com.cn/chanjing/201602/bndbu19094535shtml

The Chinese Academy of Science has started a *Design Driven Innovation*[180] by developing functional tools and visual symbols representing their intended shift to the eco-civilization in agriculture and industry. It created such tools and symbols to help managers and customers re-design processes in line with the *eco-civilization* requirements. This means a holistic and symbiotic approach rather than narrowly defined standards.

For agriculture, this means ecological high diversity and low chemical farms with high quality produce, as opposed to large monocultures depending on large amounts of agrochemicals. All this involves more responsibility for farmers. One local example of symbiotic thinking in agriculture can be found at Ying Xiang Wei Ye,[181] a farmers' corporative located in Cao Xian, Shandong, on what used to be the Yellow River's riverbed. To eliminate or minimize the use of hormones or other animal drugs, the fodder used comes from well-monitored, healthy soil, with local herbs as additives to enhance the animals' immune systems. Fresh milk delivery is limited to a prescribed distance in order to guarantee quality. Organic farms are not yet very profitable in China. The best way of improving this situation is for them to offer lodging, leisure, food and tourism in the farm. Farmers can become partners instead of employees of the enterprise.

The 13th Five-Year Plan has a strong emphasis on resource efficiency, which goes beyond the closing of inefficient companies. Club of Rome concepts like Gunter Pauli's *Blue Economy* (Sect. 3.3)[182] and *Factor Five* (Sect. 3.9)[183] have gained considerable popularity in China. Also *Cradle to Cradle*[184] and *Resource Productivity in 7 steps*[185] are available in Chinese and obtain attention. Likewise, the German concept of *Passivhaus* saving up to 90% of energy consumption became a design standard for Chinese buildings.[186] Implementing that standard would mean a tremendous shift for house construction and refurbishment in China, allowing many more people to enjoy minimized heating cost, fresh air circulation and modern LED lighting.

A very different story, also from the *Blue Economy*, is 'stone paper'[187] from sand (calcium carbonate) and plastic *waste*. It radically reduces water use, fibre from

---

[180] http://zjnews.zjol.com.cn/system/2014/10/10/020294575shtml

[181] "China Good Design", Chap. 2.3 Yin Xiang Wei Ye: Immune Health exceeding Organic Cycle, Yi Heng Cheng, China Science and Technology Publishing House, in preparation.

[182] Gunter Pauli Blue Economy, a Report to the Club of Rome, translated into Chinese by Yi Heng Cheng, Fudan University Publishing House. 2009.

[183] Ernst von Weizsäcker, Karlson Hargroves et al. Factor 5, a Report to the Club of Rome, translated into Chinese by Yi Heng Cheng, Shanghai Century Publishing, 2010.

[184] William Mc Donough and Michael Braungart, "Cradle to Cradle" translated into Chinese by 21 Century Agenda Management Company, Savage Culture Company Ltd., 2010.

[185] Michael Lettenmeier, Holger Rohn, Christa Liedtke, Friedrich Schmidt-Bleek,"Resource Productivity in 7 Steps" translated into Chinese and published by CTCI Foundation.

[186] Design Standards of Passive Low Energy Consuming Residential Buildings http://news.ces.cn/jianzhu/jianzhuzhengce/2016/01/05/98843_1.shtml

[187] "China Good Design - Green Low Carbon Innovation Design Case Study", Chap. 2.1 Stone Paper: Innovation of Nothing to Replace Something, Yi Heng Cheng et al. (2016), China Science

trees, and toxic chemicals. Stone paper can be recycled or used as an additive in steel mills, glass or cement calcination. The *Blue Economy's* philosophy actually resembles the BASF's 'Verbund' system of chemicals and energy moving through an industrial conglomerate, as much as possible using waste from one process as feedstock for the next. China has already benefitted from it, at the Yangtse BASF, Nanjing Chemical Industrial Park in Luhe Nanjing and at the Shanghai Caojing Chemical Industrial Park.

Clustering or cascading different processes can become the central methodology trait for future industrial parks. If so, industrial carbon dioxide emissions could be reduced by up to 80%, and air pollution from SOx, NOx and PM 2.5, as well as water pollution, would also be reduced. Lunan Chemical Enterprise, located in Teng Xian of Shandong, is the first coal chemical site applying coal slurry multiple injection gasification technology. Its chemical products are first and second level derivatives from methanol and ammonia.

In conclusion, it is fair to state that China's 13th Five-Year Plan to a large extent conforms to the need of globally greening the economy. China, being the biggest industrial manufacturer of the world and the model for many developing countries, seems determined to massively contribute to the greening of the world.

3.17.2 Bhutan: The Gross National Happiness Index

Until the 1970s, the sparsely populated Himalayan country of Bhutan was essentially isolated from the rest of the world. The 4th King of Bhutan, Jigme Singye Wangchuck, initiated reforms and opened the country for visitors. During the reforms, which included a modernized education system and a modernized economy, the King declared that Gross National Happiness was more important than the Gross Domestic Product, as the latter places too much emphasis on material gain over the people's well-being, bio-diversity and sustainability.

During the global financial crisis of 2008, the idea of a gross national happiness index gained a lot of traction and created excitement at the United Nations and in intellectual circles worldwide. A World Happiness Report was already being regularly published,[188] but in Bhutan the pursuit of happiness was not just a matter for philosophical discussion. For example, environmental protection is a constitutional mandate. More than 50% of Bhutan's land area is designated as protected, by national parks, nature reserves and biological corridors. The country has pledged to remain carbon neutral and to ensure that at least 60% of its landmass will remain under forest cover, in perpetuity; and in fact, the forests' carbon sequestration is currently greater than its national carbon emissions by a factor of two! Bhutan has forbidden export logging and has even instigated a monthly pedestrian day that bans all private vehicles from the roads.

---

and Technology Press. 09.2016.

[188] Helliwell et al. (2016).

Under the constitution, all Bhutanese are formally held responsible for protecting the environment, and the country is not suffering from the usual tension between economic development and environmental conservation. One fortunate geographic feature is the use of 'run-of-the-river' hydroelectricity developments, which require the preservation of watersheds in natural forests. This benign hydroelectricity meets national demands for power and still allows major exports to neighbouring India, generating a sizeable amount of foreign exchange. Bhutan has also developed a 'low impact/high value' approach to tourism, guarding against some of the negative, culturally destructive aspects of mass tourism.

In many ways, Bhutan looks pleasant and sustainable. Of course, in reality the people, especially the younger generation, also want the benefits of modern amenities. National elections in 2013 ended with a landslide victory for the People's Democratic Party, which rose from two seats to an absolute majority of 32 seats in parliament. The earlier ruling party (much focussed on 'happiness' and peace) lost 30 seats, going down from 45 to 15. The new prime minister, Thering Tobgay, has even expressed cautious scepticism about the country's famous happiness doctrine.

Today's young 5th king, Jigme Khesar Namgyel Wangchuck, however, leaves no doubt that he adheres to his father's preference for happiness over the materialistic values of the GDP. The people revere him; time will tell how Bhutan develops.

3.18 Education for a Sustainable Civilization

A consensus of leading educators is emerging who agree that radical change is needed in the global system of education, in order to meet the new and diverse needs of humanity we have been discussing.[189] Although education alone cannot achieve sustainability, it is obviously one of society's key tools. Educational objectives require a fundamental shift – from learning how to memorize and understand – to learning how to think in new, systemic ways. The real challenge is to develop in all students a capacity for problem solving, as well as critical, independent and original thinking. Education that focuses exclusively on the mind alone is no longer sufficient.[190] The radical reorientation of educational content and pedagogy should include the transmission of knowledge acquired from past experience, but should

Box: UNESCO: Education for Sustainable Development
After the UN General Assembly approved 2005–2014 as the UN Decade of Education for Sustainable Development (UNDESD), UNESCO as the responsible institution pursued the goal to integrate the principles, values and practices of sustainable development into all aspects of education. This educational

[189] Heitor Gurgulino de Souza et al., "Reflections".

[190] Zucconi (2015).

effort sought to bring about changes in behaviour among the youth and future generations to create a future that is environmentally integral, economically viable and socially just. After the completion of the UNDESD in 2014, UNESCO started a Global Action Program (GAP) to drive a worldwide effort on Education for Sustainable Development to ensure its contribution to the Sustainable Development Goals (SDGs) and the SDG4-Education 2030 Agenda. Last July, close to 100 participants of GAP Key Partners worldwide met at UNESCO in Paris, for a Monitoring Report, to be followed by a mid-term Report in 2017 and a final Report in 2019. These reports are a marker along the ESD implementation Roadmap, designed to build a better and more sustainable future for all.

also seek an expansion of the kinds of knowledge, skills and capacities that will be needed to adapt and respond creatively to a future that cannot yet be envisioned. If education is the contract between society and the future, a new contract is now needed, a contract that is no longer designed to prepare young people for a future that is largely a replica of the past. The challenge before education today is to create conditions that will enable youth to develop what the World Social Science Report (WSSR)[191] refers to as *futures literacy* – the capacity to confront complexity and uncertainty in order to dynamically participate in whatever future is encountered. In the following, some of the aspects are mentioned that seem essential for a future education system that is fit for purpose in supporting Sustainable Development.

Future education is active and collaborative. Research confirms that comprehension is lowest for passive pedagogical methods such as reading or listening to a lecture and that learning is at a maximum when it is cooperative, for example, during a discussion, group project or combined study. Eight hundred meta-analyses covering 50,000 studies on 80 million students between 2009 and 2012 revealed that cooperative learning and peer-tutoring positively affect student learning. Whereas the student's average retention rate when passively listening to a classroom lecture is 5%, practicing by doing has a 75% retention rate. It is highest, at 90%, when one student teaches another. The role of the teacher thus needs to evolve

[191] ISSC and UNESCO (2013).

Box: A Case from Napa, California

New Technology High School in Napa, California, has adopted the cooperative learning model after the city of Napa had asked a group of companies to help redesign the high school curriculum to prepare students better for career success. Based on the companies' feedback that education was essentially focused too much on individual performance rather than on the capacity to cooperate, the school redesigned the pedagogical system to centre on the person, rather than the subject. Today, it imparts not only textbook knowledge but life skills, a culture of respect, trust and responsibility. Students organize their own projects and work in groups of their choice. They are included in the decision making process in school. The curriculum is project-based and the teachers lead the activities instead of giving lectures. One criterion on which students are graded is work ethic. The School makes students help each other and see the benefit there is to be derived when competition is replaced with cooperation. The model has given rise to a global New Tech Network consisting of more than 160 schools based on the cooperative learning model.

from being a lecturer to being a guide, from imparting information to facilitating self- and peer-learning.

Future education is based on connectivity. Globally, the emerging learning model is the human network. With electronic gadgets permeating every aspect of life and learning, one often loses sight of the fact that education is essentially an organic process of exchange taking place between one human being and another. The developments in Internet and communication technology that are revolutionizing education through Massive Open Online Courses (MOOCs) and Virtual Reality training are valuable and effective only to the extent that they foster *connections between people*. Similarly, education needs to enlist the interest, release the energy and actively engage the faculties of each student to learn for oneself and also help others learn.

Future education is value-based. Values represent the quintessence of human wisdom acquired over centuries. And in the new system that's developing, they must embody the fundamental principles for sustainable accomplishment, whether individual or social. These must be even more than the inspiring ideals that supply the energy needed to fulfil human aspirations. Values are a form of knowledge and a powerful determinant of human evolution. They are psychological skills that have profound *practical* importance. Education must be founded on values that promote sustainability and general well-being for all. A move towards inculcating sustainable values would amount to a paradigm change in our current society's value system. It would consider as its aim the greater well-being of both humans and the natural systems on which they depend, rather than a valuation for more production and consumption. Conscious emphasis will be placed on values that are truly universal, as well as on respect for cultural differences. At the grass-roots level, the movement towards sustainability can build on deep local values. Values can create transformational leadership, leadership in thought that leads to action.

Future education focuses more on the topic of sustainability. As the science of sustainability is a relatively new subject, unless one is part of a traditional or indigenous culture, including it in the educational system cannot be based on centuries of work or on the collective knowledge of many past generations. Awareness of the urgency of achieving a methodology that can attain sustainable development is spreading. Not all the answers are yet available; in fact, not even all the relevant questions have been asked. Therefore, as a prior condition to education in sustainability, extensive research is needed in all branches of the subjects, including the use of multidisciplinary teams in which all interests and viewpoints are represented. Such research results need to be given wide publicity in classroom discussions and public debates that include citizens and policymakers alike. The wider and more inclusive the education and involvement of citizenry, the more effective will be the implementation.

Future education fosters an integrated way of thinking. During the last part of the twentieth century, some of the limitations of analytic thinking (see Sect. 2.7) have started to be addressed via a shift in emphasis to systems thinking. Systems thinking focuses on the interconnectedness and interdependence of phenomena and recognizes complexity while striving for a conception of the whole. However, there is still a tendency for systems thinking to view reality in relatively mechanistic terms that fail to capture its organic integrality. The limitations of systems thinking necessitate a shift from mechanistic to more organic conceptions of reality. Our greatest inventions, discoveries and acts of creativity come when apparent contradictions are reconciled. Integral thinking is thinking that is able to perceive, organize, reconcile and reunite the component elements and arrive at a truer understanding of the underlying reality – it goes beyond systems thinking in the same manner that integration goes beyond aggregation. Education must present such an integrated view to students, regardless of their specific fields of specialization. *Each discipline must likewise learn to examine itself in the light of the social whole.*

Future education fosters pluralism in content. A change in pedagogy has to be matched by a change in content. In this age of information overload and easy access to big data, choosing the right content and framing the syllabus are serious responsibilities. Social reality is complex and integrated and cannot be explained by a single theory. Many universities advocate a particular school of thought, particularly in law and economics, rather than exposing young minds to the full range of conflicting or complementary perspectives. Students today need a more inclusive form of education that seeks to complete other forms of knowledge, rather than excluding or rejecting them. One encouraging sign is the recent movement of economists and students of economics in Europe, North America and other regions, who have joined together to protest intellectual sectarianism and demand exposure to all relevant views, rather than being taught a narrow orthodoxy.[192] Just as genetic diversity has turned out to be critical for human evolution, cultural diversity can be a catalyst for social evolution. The Finnish education system (and in a similar way France) has worked on overcoming the sectoralization of curricula, as there is less of an emphasis on *subjects* and more of a focus on broader *topics*, such as the EU or ecology and space, which interrelate perspectives from many different disciplines and give students a broader overview.

---

[192] See http://www.rethinkeconomics.org/, http://reteacheconomics.org/, http://www.isipe.net/, http://www.cemus.uu.se/, http://www.schumachercollege.org.uk/, a few initiatives that necessitate a rethinking of economic theory.

Such reflections on future education become practical both at high school and university levels, for example, at the Club of Rome Schools and at North American universities:

Inspired by the former Club of Rome Report 'No Limits to Learning',[193] the German CoR Association initiated a network of *Club of Rome Schools*. Under their slogan 'Think global, act local', the 15 schools lead pupils to develop a conscience of global citizenship. Schools provide learning environments to find global perspectives and reflect on them. Curricula are mostly organized in project-learning, where students focus on a phenomenon which they explore in inter-year-groups. Education is focused on interdisciplinary skills including self-organization, self-awareness, knowledge in the field of big data and collaboration. Schools also encourage students to actively engage in relevant local projects, where they get the chance to train their self-efficacy and develop their potentials as global citizens.

McGill and York Universities in Canada, along with the University of Vermont in the United States, engage in *Education for the Anthropocene*, or Ed4A, a full graduate school system leading students to masters and doctorates on governance, law, economics and social sciences, as well as systems science and modelling, for the new professional challenges of the full world. Already a going concern, it is expanding with MOOC courses mentioned above, so as to double the current number of students (from 40 to 80) and to extend its reach beyond North America to participating universities in Australia, China and India. The programme recognizes the role of ethics and values in creating governance and economic experts who understand both natural and social systems as they currently are, not as how they have been described in the past. Club of Rome member Peter G. Brown is one of the originators of Ed4A. As he says, a new educational system is needed to cope with the new demands of sustainability research. It cannot be done 'without a complete rethink of the human project'.[194]

It is rather daunting, but the urgent need for sustainable development education does call for this kind of new paradigm. Introducing sustainability into the educational curricula at all levels is necessary; but it is also not sufficient to bring about the desired rapid, radical change in the world economy and in peoples' lifestyles. That will require raising the next generation with a different kind of training, which can impart a greater capacity for adaptation to rapid social change, as well as a strong sense of social responsibility, innovation and creative thinking. The future education system is just now in the early stages of its revolutionary transition. It will have immense impact on the future of global society as it breaks the boundaries imposed by the physical classroom, the monastic insulation of the college campus, the arbitrary rigidities of degrees, courses and 1-h lecture segments, the social barriers of class divisions and especially the economic barriers of affordability.

---

[193] J.W. Botkin, Mahdi Elmandjra, Mircea Malitza. 1979. No Limits to Learning. Bridging the Human Gap. Now as e-book at Elsevier Science Direct, 2014; CoR Schools contact: Eiken Prinz, Rosenstr.2, 20,095 Hamburg; see also http://www.club-of-rome-schulen.org/

[194] www.e4a-net.org

References

Admati A, Hellwig M (2013) The bankers new clothes. Princeton University Press, Princeton
Agarwal A, Narain S (1991) Global warming in an unequal world: a case of environmental colonialism. Centre for Science and Environment, New Delhi
Ahmed N (2016) This could be the death of the fossil fuel industry – will the rest of the economy go with it? 30 April 2016. http://www.alternet.org/environment/we-could-be-witnessing-death-fossil-fuel-industry-will-it-take-rest-economy-down-it
Arent D (2016) After Paris the smart bet is on a clean energy future. Greenmoney Journal, July/August 2016
Batty M (2013) The new science of cities. MIT Press, Cambridge, MA
Benes J, Kumhof M (2012) The Chicago plan revisited. IMF working paper 12/2012
Bowers S (2014) Luxembourg tax files. http://www.theguardian.com/business/2014/nov/05/-sp-luxembourg-tax-files-tax-avoidance-industrial-scale
Brandstetter L, Lehner OM (2015) Opening the market for impact investments: the need for adapted portfolio tools. Entrep Res J 5(2):87–107. P. 5. http://papers.ssrn.com/sol3/papers.cfm?abstract_id=2519671
Brown G (2014) Keys to building a healthy soil. You Tube. https://www.youtube.com/watch?v=9yPjoh9YJMk
Brown K, Dugan IJ (2002) Arthur Andersen's fall from grace is a sad tale of greed and miscues. Wall Street Journal, 7 June 2002. http://www.wsj.com/articles/SB1023409436545200
Carbon Tracker Initiative (2017) Expect the unexpected: the disruptive power of low-carbon technology. London
Coady D, Parry I, Sears L, Shang B (2015) How large are global energy subsidies? IMF working paper WP/15/105. https://www.imf.org/external/pubs/ft/wp/2015/wp15105.pdf
Cordell AJ, Ran Ide T, Soete L, Kamp K (1997) The new wealth of nations: taxing cyberspace. Between The Lines, Toronto. isbn:1-89637-10-5
Costanza R, dArge R, de Groot R, Farber S, Grasso M, Hannon B, Limburg K, Naeem S, Oneill RV, Paruelo J, Raskin RG, Sutton P, van den Belt M (1997) The value of the world's ecosystem services and natural capital. Nature 387(6630):253–260
Costanza R, Kubiszewski I, Giovannini E, Lovins H, McGlade J, Pickett KE, Ragnarsdóttir KV, Roberts D, Vogli RD, Wilkinson R (2014a) Time to leave GDP behind. Nature 505(7483):283–285
Costanza R, de Groot R, Sutton PC, van der Ploeg S, Anderson S, Kubiszewski I, Farber S, Turner RK (2014b) Changes in the global value of ecosystem services. Glob Environ Chang 26:152–158
Costanza R, Daly L, Fioramonti L, Giovannini E, Kubiszewski I, Mortensen LF, Pickett K, Ragnarsdóttir KV, de Vogli R, Wilkinson R (2016) Modelling and measuring sustainable well-being in connection with the UN sustainable development goals. Ecol Econ 130:350–355
Creutzig F (2015) Evolving narratives of low-carbon futures in transportation. Transp Rev 36(3):341–360
Deloitte Center for Energy Solutions (2016) The crude downturn for exploration and production companies
Dezem V, Quiroga J (2016) Chile has so much solar energy it's giving it away for free. Bloomberg, 1 June 2016
Eisler R (2007) The real wealth of nations. Creating a caring economics. Berrett Koehler, San Francisco
Erdogan B, Kant R, Miller A, Sprague K (2016) Grow fast or die slow: why unicorns are staying. Viewed on 12 May 2016 at private. http://www.mckinsey.com/industries/high-tech/our-insights/grow-fast-or-die-slow-why-unicorns-are-staying-private?
Fan R et al (2014) Anger is more influential than Joy: sentiment correlation in Weibo. doi:https://doi.org/10.1371/journal.pone.0110184
Fioramonti L (2017) The world after GDP. Polity books, Cambridge, UK

Fitzpatrick W (2016) Unlocking pension funds for impact investing. EMPEA, Legal and Regulatory Bulletin. Viewed on 15 May 2016 at http://empea.org/_files/listing_pages/UnlockingPensionFunds_Winter2016.pdf

Frey C, Kuo P (2007) Assessment of potential reductions in greenhouse gas (GHG) emissions in freight transportation. North Carolina State University, Raleigh

Fukuyama F (2001) Social capital, civil society and development. Third World Q 22(1):7–20

Fulton K, Kasper G, Kibbe B (2010) What's next for philanthropy: acting bigger and adapting better in a networked world. Monitor Institute. Downloaded 12 May 2016, 2010 from http://monitorinstitute.com/downloads/what-we-think/whats-next/Whats_Next_for_Philanthropy.pdf

GABV (2014) Global alliance for banking on values: real economy – real returns: the business case for sustainability focused banking. Full report. October 2014. Downloaded 14 May 2016 from http://www.gabv.org/wp-content/uploads/Real-Economy-Real-Returns-GABV-Research-2014.pdf

Gilding P (2015) Fossil fuels are finished – the rest is just detail. Renew Economy, 13 July 2015. http://reneweconomy.com.au/2015/fossil-fuels-are-finished-the-rest-is-just-detail-71574

Girardet H (2014) Creating regenerative cities. Routledge, Oxford

GuruFocus (2010) Gates Foundation Buys Ecolab Inc., Goldman Sachs, Monsanto Company, Exxon Mobil Corp., Sells M&T Bank, 17 Aug 2010. Viewed 10 Oct 2010 at http://www.gurufocus.com/news.php?id=104835

Hardin G (1968) The tragedy of the commons. Science 162(3859):1243–1248

Hawken P, Lovins A, Lovins H (1999) Natural capitalism: creating the next industrial revolution. Little, Brown Co, New York

Hayashi Y et al (2015) Disaster resilient city – concept and practical examples. Elsevier, Amsterdam

Helliwell J, Lay R, Sachs J (2016) World happiness report. Sustainable Development Solutions Network, New York

Henry JS (2012) The price of offshore revisited. Tax Justice Network. http://www.taxjustice.net/cms/upload/pdf/Price_of_Offshore_Revisited_120722.pdf

Higgs K (2014) Collision course: endless growth on a finite planet. MIT Press, Cambridge, MA

Höglund-Isaksson L, Sterner T (2009) Innovation effects of the Swedish NOx charge. OECD, Paris

IEA (2013) Global land transport infrastructure requirements. International Energy Agency.

ISSC and UNESCO (2013) World social science report 2013, changing global environments. OECD Publishing and UNESCO Publishing, Paris

Jamaldeen M (2016) The hidden billions. Melbourne, Oxfam, p 20

Kaldor M (2003) The idea of global civil society. Int Aff 79(3):583–593

Keane J (2003) Global civil society? Cambridge University Press, Cambridge

Kim YR (2013) A look at McKinsey & Company's biggest mistakes, 12 Sept 2013. https://www.equities.com/news/a-look-at-mckinsey-company-s-biggest-mistakes

Kohn A (1990) The brighter side of human nature: altruism and empathy in everyday life. Basic Books, New York

Koont S (2009) The urban agriculture of Havana. Mon Rev 60:44–63

Kubiszewski I (2014) Beyond GDP: are there better ways to measure well-being? The Conversation, 2 Dec 2014. http://theconversation.com/beyond-gdp-are-there-better-ways-to-measure-well-being-33414

Kubiszewski I, Costanza R, Franco C, Lawn P, Talberth J, Jackson T, Aylmer C (2013) Beyond GDP: measuring and achieving global genuine progress. Ecol Econ 93:57–68

Kuenkel P (2016) The art of collective leadership. Chelsea Green, White River Junction

Kuenkel P et al (2009) The common code for the coffee community. In: Volmer D (ed) Enhancing the effectiveness of sustainability partnerships. National Academies Press, Washington, DC

Kuipers D (2015) Buying the farm. Orion, July 2015. https://orionmagazine.org/article/buying-the-farm/

Lovins A, Rocky Mountain Institute (2011) Reinventing fire. Bold business solutions for the new energy era. Chelsea Green, White River Junction, p xi

Mankiw NG (1998) Principles of economics. The Dryden Press, Fort Worth

Mann A, Harter J (2016). Worldwide employee engagement crisis. Gallup Business Journal, 7 Jan 2016
Maxton G, Randers J (2016) Reinventing prosperity. Greystone Books, Vancouver/Berkeley
McKinsey (2017) Basel IV: what's next for banks? Global Risk Practice, Apr 2017
Meadows D, Randers J, Meadows D (1992) Beyond the limits. Confronting global collapse. Chelsea Green, White River Junction
Moore CJ, Moore SL, Weisberg SB, Lattin GL, Zellers AF (2001). A comparison of neustonic plastic and zooplankton abundance in southern California's coastal waters
Newman P, Kenworthy J (1989) Cities and automobile dependence: an international sourcebook. Gower, Aldershot
Newman P, Kenworthy J (2015) The end of automobile dependence. Island Press, Washington
OECD (2016) New steps to strengthen transparency in international tax matters. http://www.oecd.org/tax/automatic-exchange/news/new-steps-to-strengthen-transparency-in-international-tax-matters-oecd-releases-guidance-on-the-implementation-of-country-by-country-reporting.htm
OECD (2017) Mobilising bond markets for a low-carbon transition. OECD, Paris
Ogg JC (2016) Ahead of model 3: tesla value for 2019 versus Ford and GM Today. 24/7 Wall St, 29 Mar 2016. http://247wallst.com/autos/2016/03/29/ahead-of-model-3-tesla-value-for-2019-versus-ford-and-gm-today/
Pauli G (2010) The blue economy. Ten years, 100 innovations, 100 million jobs. A report to the Club of Rome
Pauli G (2015) The blue economy version 2.0. 200 projects implemented, US$ 4 billion invested, 3 million jobs created. Academic Foundation, New Delhi and London
Piketty T (2015) La dette publique est. une blague! Interview by Reporterre viewed 14 May 2016 at http://pressformore.com/view/la-dette-publique-est-une-blague-la-vraie-dette-est-celle-du-capital-naturel
Piller C, Sanders E, Dixon R (2007) Dark clouds over good works of Gates Foundation. Retrieved 3 Oct 2010, from latimes.com/news/nationworld/nation/la-na-gatesx07jan07,0,6,827,615.story
Pledge G (2010) Forty U.S. families take giving pledge: Billionaires pledge majority of wealth to philanthropy. Retrieved 10 Sept 2010, from http://givingpledge.org/Content/media/PressRelease_8_4.pdf
Pyper J (2012) To boost gas mileage, automakers explore lighter cars. Scientific American
Radford T (2015) Stop burning fossil fuels now: there is no CO2 technofix. The Guardian, 2 Aug 2015
Randall T (2015) Fossil fuels just lost the race against renewables. Bloomberg Business, 14 Apr 2015. http://www.bloomberg.com/news/articles/2015-04-14/fossil-fuels-just-lost-the-race-against-renewables
Raskin P (2016) Journey to Earthland. The great transition to planetary civilization. Tellus Institute, Boston
Raworth K (2017) Doughnut economics. Penguin Random House, London
RMI (2011) Reinventing fire: transportation sector methodology. Rocky Mountain Institute, USA
Rockström J et al (2017) A roadmap for rapid decarbonisation. Science 355:1269–1271
Sanders R (2006) Sustainability – implications for growth, employment and consumption. International Journal of Environment, Workplace and Employment, Jan. 2006
Seba T (2014) Clean disruption of energy and transportation: how silicon valley will make oil, nuclear, natural gas, coal, electric utilities and conventional cars obsolete by 2030, 20 May 2014. https://www.amazon.com/Clean-Disruption-Energy-Transportation-Conventional/dp/0692210539?ie=UTF8&redirect=true
Shaxson N (2012) Treasure islands: tax havens and the men who stole the world. Vintage, London
Snowden DJ, Boone M (2007) A leader's framework for decision making. Harv Bus Rev:69–76
Spratt D, Sutton P (2008) Climate code red: the case for emergency action. Scribe Publications, Brunswick
SRI-Rice (2014) The system of crop intensification. SRI International Network and Resources Center, Cornell University, and the Technical Centre for Agricultural and Rural Cooperation (CTA), Wageningen

Steffen A (2015) A talk given at a conservation meeting a hundred years from now. 3 Nov 2015. http://www.alexsteffen.com/future_conservation_meeting_talk
Stuchtey M, Enkvist P-A, Zumwinkel K (2016) A good disruption redefining growth in the twenty-first century. Bloomsbury Publishing Plc, London
Tartar A (2016) World War II economy is a master class in how to fight climate change. Bloomberg Markets 8 Sep 2016
Tekelova M (2012) Full-reserve banking, a few simple changes to banking that could end the debt crisis. Positive Money, London
The Club of Rome (2015) The circular economy and benefits for society Swedish case study shows jobs and climate as clear winners. Club of Rome, Winterthur
The International Centre of Insect Physiology and Ecology (ICIPE) (2015) The 'push–pull' farming system: climate-smart, sustainable agriculture for Africa.
Thorndike J (2017) Refundable carbon tax – not perfect but good enough. Forbes. 19 Feb 2017
Tilly C (2004) Social movements, 1768–2004. Routledge, Oxford
Turner A (2016) Between debt and the devil: money, credit and fixing global finance. Princeton University Press, Princeton
UN PRI (2013) Overcoming strategic barriers to a sustainable financial system: a consultation with signatories on a new PRI work programme. Viewed 13 May 2016 at https://www.unpri.org/explore/?q=Overcoming+strategic+barriers+to+a+sustainable+financial+system
UNEP (2016), Food Systems and Natural Resources (2016). By Henk Westhoek et al. UNEP, Nairobi
UNEP and International Resource Panel (2014) Assessing global land use: balancing consumption with sustainable supply. UNEP, Nairobi
Uphadhyay A (2016) Narendra Modi lures India's top fossil fuel companies to back solar boom. Live Mint, 22 July 2016. http://www.livemint.com/Industry/n6JGIUiAK3dBZHvUxWprWO/Narendra-Modi-lures-Indias-top-fossil-fuel-companies-to-bac.html
Uphoff N (2008) The system of rice intensification (SRI). Jurnal Tanah dan Lingkungan 10(1):27–40
Vaughn A (2016) Human impact has pushed Earth into the Anthropocene, scientists say. The Guardian, 7 Jan 2016. http://www.theguardian.com/environment/2016/jan/07/human-impact-has-pushed-earth-into-the-anthropocene-scientists-say
von Weizsäcker E, Jesinghaus J (1992) Ecological tax reform. Zed Books, London
von Weizsäcker E, "Charlie" Hargroves K et al (2009) Factor five: transforming the global economy by 80% improvements in resource productivity. Earthscan, London
WBGU (2009) Solving the climate dilemma: the budget approach. WBGU, Berlin
WBGU (2016) Humanity on the move: unlocking the transformative power of *cities*. German Advisory Council on Global Change, Berlin
West M (2016a) Oligarchs of the Treasure Islands. Interview with George Rozvany. 11 July 2016. http://www.michaelwest.com.au/oligarchs-of-the-treasure-islands/
West M (2016b) George Rozvany's silver bullet. 13 July 2016. http://www.michaelwest.com.au/george-rozvanys-silver-bullet/
Wolfensen K (2013) Coping with the food and agriculture challenge: smallholders' agenda. FAO, p 1. http://www.fao.org/fileadmin/templates/nr/sustainability_pathways/docs/Coping_with_food_and_agriculture_challenge__Smallholder_s_agenda_Final.pdf
Wolman A (1965) The metabolism of cities. Sci Am 213:179–190
World Future Council Foundation (2014) Global policy action plan. Incentives for a sustainable future. Oeding Print, Braunschweig. For more go to www.worldfuturecouncil.org or mail to the program's coordinator, catherine.pearce@worldfuturecouncil.org
Zucconi A (2015) Person centered education. Cadmus 2(5):59–61
Zucman G (2015) The hidden wealth of nations: the scourge of tax havens (transl. Fagan T). University of Chicago Press, Chicago. Data available here: www.gabriel-zucman.eu

Conclusion: We Invite Readers to 'Come on'

We as authors invite readers to join us on an exciting journey. Chapter 3 of this book is full of examples showing that some courageous individuals or companies or states can act now, leaving misery, frustration and stagnation behind. Chapter 3 also shows that policy measures exist or are conceivable to make constructive acts profitable and lift them into the mainstream.

Our invitation is addressed to readers in all countries of the world. Our conditions differ widely, but the world is now witnessing success stories popping up in the most unlikely places.

One condition should be respected everywhere: leaving the trajectory of unsustainable growth. It is unfair to future generations. It is particularly unfair to the poorest in this world to continue 'mining' the environment, destroying biodiversity and destabilizing climate. The poor depend on the surrounding local environment and on a reasonably stable climate, as do all the other living beings with which we share the planet.

The current trends on Earth are not sustainable and the usual answers to these challenges are usually dependent on the kind of economic growth that is firmly concatenated with additional resource consumption. Combined with a continued population increase, this makes current trends even less sustainable. The unavoidable result of this is local and global ecological collapses which will, along with many lives, completely destroy the Sustainable Development Goals (SDGs) 13, 14 and 15. It seems unavoidable that what is urgently required is the development of new kinds of human goals, and if possible, a new social Enlightenment. One of the main characteristics of that enlightenment is *balance*. The goal is a balanced world, with a realistic harmony between the current list of economic and ecological SGDs.

We invite engineers, inventors, practitioners and financial investors to work on decoupling economic success and human satisfaction from the consumption of natural resources. That programme includes the recovery of used resources. It also includes the restoration of degraded land so as to improve conditions for wildlife and for fecund agriculture.

We invite families, notably in countries with continued population increases, to aim for a stabilization in the human population, and we urge states to establish and maintain social security for families, independent of their number of children.

© Springer Science+Business Media LLC 2018

E.U. von Weizsäcker, A. Wijkman, *Come On!*,
DOI 10.1007/978-1-4939-7419-1

We invite academic readers to help overhaul the mechanistic and materialistic philosophy that often shows an abundance of mathematics and a vacuity of meaning. We encourage states and private sponsors of academic institutions to support interdisciplinary research and academic degrees.

We invite business people to honour the common good and the long-term perspective, as opposed to the quarterly reports of superficial financial success. That will require the financial community getting *off* the gravy train, and becoming more patient and realistic about reasonable profit margins.

We invite the business community to keep in touch with policymakers, with a view of changing the basic frameworks for profit-making so that contributions to the common good are financially rewarded, not penalized. We also suggest rewards (and awards) for exemplary behaviour.

We invite policymakers to introduce a new philosophy of taxation that rewards employing people and exacts penalties for the consumption of natural resources, while continuing to respect everyone's need to have affordable access to necessary resources.

We invite governments to come together across borders to cooperate on common advantages in the spirit of 'cohabitation'.

Last but not least, we invite critics to spell out what they find wrong in terms of the facts and intentions of this Report from the Club of Rome.

Praise for *Come On!*

Dr. Jeremy Leggett: Founder and Chairman, Solarcentury & SolarAid Chairman, Carbon Tracker
The Club of Rome has understood the power of our divestment campaign that turns out to be more powerful than politicians wanting the coal era return. Welcome back on the international arena, dear Club of Rome!

Prof. Kevin Chika Urama African Development Bank, Abidjan
The Club of Rome is no longer talking about stopping growth. It gets even bolder, suggesting a New Enlightenment overcoming Western/Northern development pathways which has been viewed by many as the panacea for improved wellbeing. The book is a must read for Development Practitioners and Policymakers in the developing countries, especially in Africa.

Dr. Janez Potocnik, Former EU Environment Commissioner, Co-Chair, International Resource Panel
'Limits of Growth' might not have been exact in all details, but the core message was always alive and is now more topical than ever. But our civilization needs a New Enlightenment to cope with the challenges of the 'Full World'. The new Report by Club of Rome is here to provide you exactly that (easy to read, simple to understand, difficult to deny, impossible to ignore).

Ms. Mizue Tsukushi, Founder and President, The Good Bankers, Japan
The new Report from the Club of Rome puts great emphasis on benign investments. Bravo. Hope found. This is also our philosophy at The Good Bankers in Japan.

Prof. Lu Yonglong, Chinese Academy of Sciences
The new Club of Rome Report presents not only an excellent account of China's 13th Five Year Plan but a 'deepening reform' choice for an emerging economy to develop a green society. It also covers the areas of great significance for the transformation towards a sustainable world.

© Springer Science+Business Media LLC 2018
E.U. von Weizsäcker, A. Wijkman, *Come On!*,
DOI 10.1007/978-1-4939-7419-1

Prof. James Gustave Speth, Founder, World Resources Institute, Former Administrator, UN Development Programme
A powerful new Club of Rome Report! The authors are right saying that the 'Full World' needs the equivalent of a New Enlightenment. Best of all, the book is optimistic and concrete in what can be done now!

Prof. Mark Swilling, University of Stellenbosch, South Africa
The new Club of Rome Report courageously proposes a New Enlightenment for our 'Full World', and highlights the need for a just transition through a systems-based approach. The report also offers a powerful critique of the brutal financialization of the world economy.

Sunita Narain, Director, Centre for Science and Environment, New Delhi
The Club of Rome comes with a powerful new report. It suggests a new Enlightenment, one fitting today's 'Full World'. It also presents success stories and bold policy proposals that are worth considering.

Prof. Johan Rockström, Director, Stockholm Resilience Institute
'The scientific message is clear. The world faces an imminent social transformation to global sustainability. The only chance for humanity to prosper through the twenty-first century is within the limits of the planet, or in other words – within the safe operating space of planetary boundaries. This new Club of Rome Report is a critical state-of-the-art scan of the necessity, opportunity and benefits of a world transition to sustainable development. A book we all need'.

Kate Raworth, Oxford University, Author, Doughnut Economics
von Weizsäcker and Wijkman are right: it's time for a new enlightenment if humanity is to thrive this century – and yet we cannot wait for enlightenment before taking decisive action. This rich and timely book draws on a wide array of innovative thinkers to make a powerful case for action today that leads towards a future of thriving balance for all.

Victoria Tauli Corpuz, The Philippines, UN Special Rapporteur on the Rights of Indigenous Peoples
The report carries optimism in an otherwise difficult world. I like the idea of a balance-oriented New Enlightenment. Asian Cultures know its power better than the capitalist and colonialist countries of the West.

Prof. Klaus Töpfer, Former Executive Director, UNEP, Nairobi
Forty-five years after *The Limits to Growth* a new Report from the Club of Rome; and a courageous one and full of optimism! It goes as far as proposing a New Enlightenment, appropriate for the Anthropocene.

Gilberto Gallopín, Co-author of the Latin American World Model
As one of the Latin American critics of the first report of the Club of Rome, I welcome the change in focus of its new report towards tackling the fundamental root causes and potential solutions of the planetary predicament. The report shows convincingly that a New Enlightenment is not only necessary but also possible, and makes a call to action.

Diana Schumacher, Former Director, New Economics Foundation, Godstone, UK
Finally, we have another big report from the Club of Rome, 45 years after *The Limits to Growth*. But this time the authors go further and deeper. They suggest that the 'Full World' needs a new Enlightenment beyond materialism, reductionism and selfishness. I do hope their call will be heard!

Connie Hedegaard, Former EU Commissioner for Climate
The new report from the Club of Rome comes timely. Time is running out and we must transform all major sectors of the economy to stay within the planetary boundaries. This will require a more systemic approach and, indeed, priority given to the long term. The values guiding us have to be broadened and I believe the authors are right in their plea for a new Enlightenment.

We have created an email for reader's response: comeonauthors@clubofrome.org

Index

© Springer Science+Business Media LLC 2018
E.U. von Weizsäcker, A. Wijkman, *Come On!*,
DOI 10.1007/978-1-4939-7419-1

D

E